American Book Company
The Standards Experts

California Integrated Mathematics 1

ERICA DAY

COLLEEN PINTOZZI

AMERICAN BOOK COMPANY

P. O. BOX 2638

WOODSTOCK, GEORGIA 30188-1383

TOLL FREE 1 (888) 264-5877   PHONE (770) 928-2834

TOLL FREE FAX 1 (866) 827-3240

WEB SITE: www.americanbookcompany.com

Acknowledgements

In preparing this book, we would like to acknowledge Samuel Rodriguez for his contribution in writing and editing and Mary Stoddard and Eric Field for their contributions developing graphics for this book. We would also like to thank our many students whose needs and questions inspired us to write this text.

Copyright © 2007
by American Book Company
P.O. Box 2638
Woodstock, GA 30188-1383

ALL RIGHTS RESERVED

The text of this publication, or any part thereof, may not be reproduced or transmitted in any form or by any means, electronic or mechanical, including photocopying, recording, storage in an information retrieval system, or otherwise, without the prior permission of the publisher.

Printed in the United States of America
02/09   06/07

Contents

Acknowledgements		ii
Preface		viii
Diagnostic Test		1
1 Exponents		12
1.1	Understanding Rational Numbers	12
1.2	Opposite Numbers	13
1.3	Reciprocals	14
1.4	Understanding Exponents	14
1.5	Multiplying Exponents with the Same Base	15
1.6	Multiplying Fractional Exponents with the Same Base	16
1.7	Multiplying Exponents Raised to an Exponent	16
1.8	Expressions Raised to a Power	17
1.9	Fractions Raised to a Power	17
1.10	More Multiplying Exponents	18
1.11	More Multiplying Fractional Exponents	18
1.12	Negative Exponents	19
1.13	Multiplying with Negative Exponents	19
1.14	Dividing with Exponents	20
1.15	Dividing with Fractional Exponents	21
1.16	Order of Operations	22
	Chapter 1 Review	24
	Chapter 1 Test	25
2 Roots		27
2.1	Square Root	27
2.2	Simplifying Square Roots	27
2.3	Adding and Subtracting Roots	28
2.4	Multiplying Roots	29
2.5	Dividing Roots	30
2.6	Cube Roots	31
2.7	Fractional Exponents	31
	Chapter 2 Review	32
	Chapter 2 Test	33

3 Introduction to Algebra — 34

- 3.1 Algebra Vocabulary — 34
- 3.2 Substituting Numbers for Variables — 35
- 3.3 Understanding Algebra Word Problems — 36
- 3.4 Setting Up Algebra Word Problems — 38
- 3.5 Changing Algebra Word Problems to Algebraic Equations — 39
- 3.6 Substituting Numbers in Formulas — 40
- 3.7 Properties of Addition and Multiplication — 41
- Chapter 3 Review — 42
- Chapter 3 Test — 44

4 Solving Multi-Step Equations and Inequalities — 46

- 4.1 Two-Step Algebra Problems — 46
- 4.2 Two-Step Algebra Problems with Fractions — 47
- 4.3 More Two-Step Algebra Problems with Fractions — 48
- 4.4 Combining Like Terms — 49
- 4.5 Solving Equations with Like Terms — 49
- 4.6 Removing Parentheses — 51
- 4.7 Multi-Step Algebra Problems — 52
- 4.8 Multi-Step Inequalities — 54
- Chapter 4 Review — 56
- Chapter 4 Test — 57

5 Algebra Word Problems — 58

- 5.1 Geometry Word Problems — 59
- 5.2 Age Problems — 60
- 5.3 Consecutive Integer Problems — 62
- 5.4 Inequality Word Problems — 63
- Chapter 5 Review — 64
- Chapter 5 Test — 65

6 Polynomials — 66

- 6.1 Adding and Subtracting Monomials — 66
- 6.2 Adding Polynomials — 67
- 6.3 Subtracting Polynomials — 68
- 6.4 Multiplying Monomials — 70
- 6.5 Multiplying Monomials with Different Variables — 71
- 6.6 Dividing Monomials — 72

Contents

	6.7	Multiplying Monomials by Polynomials	73
	6.8	Dividing Polynomials by Monomials	74
	6.9	Removing Parentheses and Simplifying	75
	6.10	Multiplying Two Binomials	76
	6.11	Simplifying Expressions with Exponents	78
		Chapter 6 Review	79
		Chapter 6 Test	80

7 Factoring — **82**
- 7.1 Finding the Greatest Common Factor of Polynomials — 82
- 7.2 Factor By Grouping — 85
- 7.3 Factoring Trinomials — 86
- 7.4 More Factoring Trinomials — 88
- 7.5 Factoring More Trinomials — 89
- 7.6 Factoring the Difference of Two Squares — 91
- 7.7 Simplifying Rational Expressions — 93
- 7.8 Adding Rational Expressions — 94
- 7.9 Subtracting Rational Expressions — 95
- 7.10 Multiplying Rational Expressions — 96
- 7.11 Dividing Rational Expressions — 97
- Chapter 7 Review — 98
- Chapter 7 Test — 99

8 Solving Quadratic Equations — **101**
- 8.1 Solving the Difference of Two Squares — 103
- 8.2 Solving Perfect Squares — 105
- 8.3 Completing the Square — 106
- 8.4 Proof of the Quadratic Formula — 107
- 8.5 Using the Quadratic Formula — 108
- Chapter 8 Review — 109
- Chapter 8 Test — 110

9 Graphing and Writing Equations and Inequalities — **111**
- 9.1 Forms of Linear Equations — 111
- 9.2 Graphing Linear Equations — 112
- 9.3 Graphing Horizontal and Vertical Lines — 114
- 9.4 Finding the Distance Between Two Points — 115
- 9.5 Finding the Midpoint of a Line Segment — 116

9.6	Finding the Intercepts of a Line	117
9.7	Understanding Slope	118
9.8	Slope-Intercept Form of a Line	120
9.9	Verify That a Point Lies on a Line	121
9.10	Graphing a Line Knowing a Point and Slope	122
9.11	Finding the Equation of a Line Using Two Points or a Point and Slope	123
9.12	Equations of Parallel Lines	124
9.13	Equations of Perpendicular Lines	124
9.14	Graphing Inequalities	127
	Chapter 9 Review	130
	Chapter 9 Test	131

10 Mathematical Reasoning — 135

10.1	Mathematical Reasoning/Logic	135
10.2	Deductive and Inductive Arguments	137
	Chapter 10 Review	139
	Chapter 10 Test	140

11 Triangles — 141

11.1	Types of Triangles	141
11.2	Interior Angles of a Triangle	142
11.3	Exterior Angles	143
11.4	Triangle Inequality Theorem	144
	Chapter 11 Review	145
	Chapter 11 Test	146

12 Plane Geometry — 147

12.1	Types of Polygons	147
12.2	Sum of Interior Angles of a Polygon	148
12.3	Perimeter	149
12.4	Area of Squares and Rectangles	150
12.5	Area of Triangles	151
12.6	Area of Trapezoids and Parallelograms	152
12.7	Area of a Rhombus	153
12.8	Circumference	154
12.9	Area of a Circle	155
12.10	Two-Step Area Problems	156
12.11	Geometric Relationships of Plane Figures	158

Contents

    Chapter 12 Review   160
    Chapter 12 Test   161

13 Solid Geometry   **162**
  13.1 Understanding Volume   162
  13.2 Volume of Rectangular Prisms and Cubes   163
  13.3 Volume of Spheres, Cones, Cylinders, and Pyramids   164
  13.4 Two-Step Volume Problems   166
  13.5 Geometric Relationships of Solids   167
  13.6 Surface Area   169
  13.7 Cube   169
  13.8 Rectangular Prisms   169
  13.9 Pyramid   171
  13.10 Cylinder   172
  13.11 Sphere   173
  13.12 Cone   173
  13.13 Solid Geometry Word Problems   174
    Chapter 13 Review   175
    Chapter 13 Test   177

Practice Test 1   **179**

Practice Test 2   **188**

Index   197

Preface

California Integrated Math 1 will help you review and learn important concepts and skills related to high school mathematics. To help identify which areas are of greater challenge for you, first take the diagnostic test, then complete the evaluation chart with your instructor in order to help you identify the chapters which require your careful attention. When you have finished your review of all of the material your teacher assigns, take the practice tests to evaluate your understanding of the material presented in this book. **The materials in this book are based on the standards in mathematics published by the California Department of Education. The complete list of standards is located on the next page and at the beginning of the Answer Key. Each question in the Diagnostic and Practice Tests is referenced to the standard, as is the beginning of each chapter.**

This book contains several sections. These sections are as follows: 1) A Diagnostic Test; 2) Chapters that teach the concepts and skills for the CA Integrated Mathematics I course; 3) Two Practice Tests. Answers to the tests and exercises are in a separate manual.

ABOUT THE AUTHORS

Erica Day has a Bachelor of Science Degree in Mathematics and working on a Master of Science Degree in Mathematics. She graduated with high honors from Kennesaw State University in Kennesaw, Georgia. She has also tutored all levels of mathematics, ranging from high school algebra and geometry to university-level statistics, calculus, and linear algebra. She is currently writing and editing mathematics books for American Book Company, where she has coauthored numerous books, such as ***Passing the Georgia Algebra I End of Course, Passing the Georgia High School Graduation Test in Mathematics, Passing the Arizona AIMS in Mathematics,*** and ***Passing the New Jersey HSPA in Mathematics***, to help students pass graduation and end of course exams.

Colleen Pintozzi has taught mathematics at the middle school, junior high, senior high, and adult level for 22 years. She holds a B.S. degree from Wright State University in Dayton, Ohio and has done graduate work at Wright State University, Duke University, and the University of North Carolina at Chapel Hill. She is the author of many mathematics books including such best-sellers as ***Basics Made Easy: Mathematics Review, Passing the New Alabama Graduation Exam in Mathematics, Passing the Louisiana LEAP 21 GEE, Passing the Indiana ISTEP+ GQE in Mathematics, Passing the Minnesota Basic Standards Test in Mathematics,*** and ***Passing the Nevada High School Proficiency Exam in Mathematics.***

ABOUT THE REVIEWER

D'Lane McMillin has a Bachelor of Science Degree in Chemistry and a Master of Science Degree in Chemistry/Science Education from SW Oklahoma State University. She has taught chemistry at the junior college level for 5 years, mathematics at the middle school level for 8 years, and mathematics and chemistry at the high school level for 5 years. She is currently the K-12 District Math Curriculum Coordinator for Desert Sands USD. She has also been a presenter for curriculum math mapping workshops (K-12) and class management, lesson planning, and math instructional strategies seminars.

California Content Standards for Integrated Math 1

	Algebra I (A)	# of test questions
1.1	Students use properties of numbers to demonstrate whether assertions are true or false.	1/2
2.0	Students understand and use such operations as taking the opposite, finding the reciprocal, taking a roots, and raising to a fractional power.	4
4.0	Students simplify expressions prior to solving linear equations and inequalities in one variable such as $3(2x-5)+4(x-2)=12$.	3
5.0	Students solve multistep problems, including word problems, involving linear equations and linear inequalities in one variable and provide justification for each step.	6
6.0	Students graph a linear equation, and compute the x- and y-intercepts (e.g., graph $2x+6y=4$). They are also able to sketch the region defined by linear inequality (e.g., they sketch the region defined by $2x+6y<4$).	4
7.0	Students verify that a point lies on a line given an equation of the line. Students are able to derive linear equations by using the point-slope formula.	4
8.0	Students understand the concepts of parallel lines and perpendicular lines and how those slopes are related. Students are able to find the equation of a line perpendicular to a given line that passes through a given point.	1
10.0	Students add, subtract, multiply, and divide monomials and polynomials. Students solve multistep problems, including word problems, by using these techniques.	4
11.0	Students apply basic factoring techniques to second- and simple third-degree polynomials. These techniques include finding a common factor for all terms in a polynomial, recognizing the difference of two squares, and recognizing perfect square of binomials.	2
12.0	Students simplify fractions with polynomials in the numerator and denominator by factoring both and reducing them to lowest terms.	3
13.0	Students add, subtract, multiply, and divide rational expressions and functions. Students solve both computationally and conceptually challenging problems by using these techniques.	4
14.0	Students solve a quadratic equations by factoring or completing the square.	3
19.0	Students know the quadratic formula and are familiar with its proof by completing the square.	2
20.0	Students use the quadratic formula to find the roots of a second-degree polynomial and to solve quadratic equations.	3
25.1	Students use properties of numbers to construct simple, valid arguments (direct and indirect) for, or formulate counterexamples to, claimed assertions.	1/2
25.2	Students judge the validity of an argument according to whether the properties of the real number system and the order of operations have been applied correctly at each step.	1/2
25.3	Given a specific algebraic statement involving linear, quadratic, or absolute value expressions or equations or inequalities, students determine whether the statement is true sometimes, always, or never.	1/2

Preface

Geometry (G) | # of test questions

6.0	Students know and are able to use the triangle inequality theorem.	1
8.0	Students know, derive, and solve problems involving perimeter, circumference, area, volume, lateral area, and surface area of common geometric figures.	4
9.0	Students compute the volumes and surface areas of prisms, pyramids, cylinders, cones, and spheres; and students commit to memory the formulas for prisms, pyramids, and cylinders.	2
10.0	Students compute areas of polygons, including rectangles, scalene triangles, equilateral triangles, rhombi, parallelograms, and trapezoids.	4
11.0	Students determine how changes in dimensions affect the perimeter, area, and volume of common geometric figures and solids.	1
12.0	Students find and use measures of sides and of interior and exterior angles of triangles and polygons to classify figures and solve problems.	5
17.0	Students prove theorems by using coordinate geometry, including the midpoint of a line segment, the distance formula, and various forms of equations of lines and circles.	3
18.0	Students know the definitions of the basic trigonometric functions defined by the angles of a right triangle. They also know and are able to use elementary relationships between them. For example, $\tan(x) = \sin(x)/\cos(x)$, $(\sin(x))^2 + (\cos(x))^2 = 1$.	0

Integrated Math 1 Total   **65**

Diagnostic Test

1. Simplify the following expression: $\dfrac{5^4}{5^2}$

   (A) 5^2
   (B) 5^6
   (C) 5^8
   (D) 5^{42}

   A2.0

2. Antoine has a circular pool that has a radius of 4 feet. Kirk has a circular pool that has a radius of 8 feet. How many times greater is the surface area of the top of Kirk's pool compared with the top Antoine's pool?

   (A) 2 times greater
   (B) 3 times greater
   (C) 4 times greater
   (D) 8 times greater

   G11.0

3. For the following question, use the formula $V = l \times w \times h$

   Jim has an aquarium with a length of 12 inches, a width of 12 inches, and a height of 18 inches. What is the volume of his aquarium?

   (A) $2,592\pi$ in^3
   (B) $2,592$ in^3
   (C) $1,296$ in^3
   (D) 144 in^3

   G9.0

4. Find the equation of the line that has a slope of $-\frac{1}{2}$ and runs through point $(2, 1)$.

   (A) $y = -\frac{1}{2}x + 1$
   (B) $y = -\frac{1}{2}x + 2$
   (C) $y = 2x + 2$
   (D) $y = \frac{1}{2}x + 1$

   A7.0

5. There were three brothers, Fernando was two years older than Pedro. Pedro was two years older than Samuel. Together their ages add up to 63 years. How old is Samuel?

   (A) 17
   (B) 19
   (C) 21
   (D) 23

   A5.0

6. Use the following diagram for this question:

   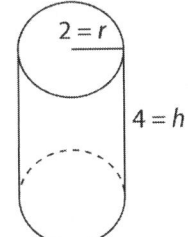

   What is the volume of the cylinder above?

   (A) 16
   (B) 8π
   (C) 16π
   (D) 32π

   G9.0

7. Which of the following points is NOT found on the line $y = -7x - 6$?

   $Q: (1, -13), R: (6, -36), S: (-2, -20)$

   (A) R
   (B) Q, R
   (C) Q, S
   (D) R, S

   A7.0

Copyright © American Book Company

8. Use the following diagram for this question:

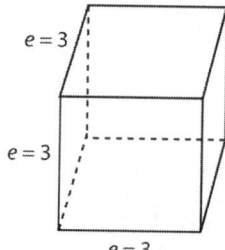

Which of the following is the surface area of a cube with side 3 m?

(A) 9 m²
(B) 27 m³
(C) 54 m²
(D) 81 m³

G8.0

9. What is the equation of a line parallel to the y-axis and four units to the left of the y-axis in the xy plane?

(A) $x = 4$
(B) $x = -4$
(C) $y = 4$
(D) $y = -4$

A8.0

10. An automobile manufacturer produces two kinds of cars, the x and the y. The company must always produce at least 200 cars but no more than 1,000 cars per day. Which of the following represents this situation?

(A) $200 \leq x + y \geq 1,000$
(B) $200 \leq x \leq 1,000$
(C) $200 \leq x + y \leq 1,000$
(D) None of the Above

A5.0

11. Use the following diagram to answer this question.

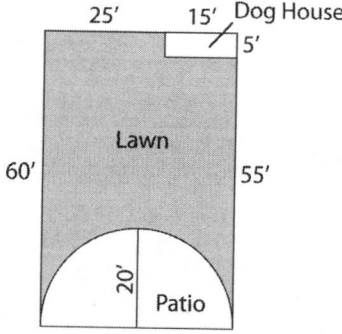

Rodney needs sod to cover his new lawn. He has a 5 foot by 15 foot section for his doghouse, and his circular patio extends into the lawn in an exact semi-circle with radius 20 feet. From the diagram, calculate about how much sod Rodney will need in ft².

(A) $2,400 - 200\pi$ ft²
(B) $2,400 - 400\pi$ ft²
(C) $2,325 - 200\pi$ ft²
(D) $2,325 - 400\pi$ ft²

G8.0

12. Simplify: $\dfrac{x^2 - 2x - 15}{x^2 + x - 6}$, if $x \neq 2$ or $x \neq 3$.

(A) $\dfrac{x-5}{x+2}$
(B) $\dfrac{x-5}{x-2}$
(C) $\dfrac{x+5}{x-2}$
(D) $\dfrac{x+5}{x+2}$

A12.0

13. Write an equation of a line that has an undefined slope and passes through the point $(3, -4)$.

(A) $y = 3$
(B) $x = 3$
(C) $y = -4$
(D) $x = -4$

A7.0

14. Lauren had a box of 48 candy bars to sell for a club fund-raiser. She sold half of the bars on her own, and her father sold half of the remaining bars at work. If no other bars were sold, what fraction of Lauren's original bars were sold?

(A) $\frac{1}{3}$

(B) $\frac{1}{4}$

(C) $\frac{2}{3}$

(D) $\frac{3}{4}$

A5.0

15. Which of the following is a graph of the inequality $y \leq x - 3$?

(A)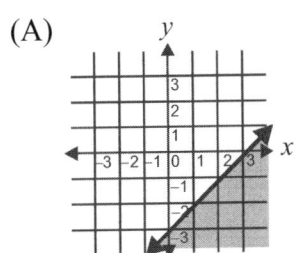

(B)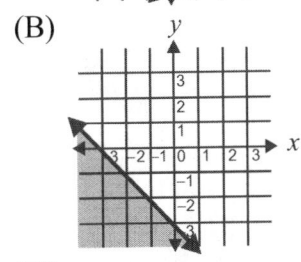

(C)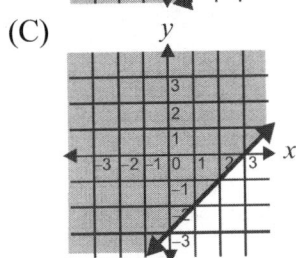

(D)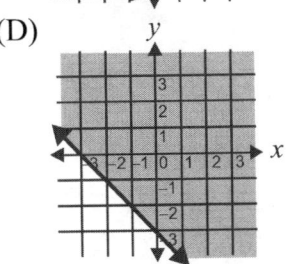

A6.0

16. What are the coordinates of the midpoint of $\overline{AB}$ if $A = (0, -6)$ and $B = (-2, 1)$?

(A) $(-2, -5)$
(B) $(-1, -3.5)$
(C) $(1, -3.5)$
(D) $(-1, -2.5)$

G17.0

17. Which of the following is equivalent to 3^{-5}?

(A) -45
(B) -15
(C) $\frac{1}{243}$
(D) 3×10^{-5}

A2.0

18. Isabella is simplifying this expression:
$2(5a + 3b - c) - 5(4a - 2b - 3c)$
The expression above is equivalent to which of the following expressions?

(A) $-10a + 16b + 13c$
(B) $-10a - 4b - 4c$
(C) $30a + b + 2c$
(D) $30a - 4b - 17c$

A10.0

19. What is the x-intercept of the following linear equation?
$3x + 4y = 12$

(A) $(0, 3)$
(B) $(3, 0)$
(C) $(0, 4)$
(D) $(4, 0)$

A6.0

20. Solve: $6 - 2(5y - 1) = 18$

(A) $y = 2$
(B) $y = -2$
(C) $y = 1$
(D) $y = -1$

A4.0

21. Simplify: $\sqrt{45} \times \sqrt{27}$

   (A) $3\sqrt{15}$
   (B) $9\sqrt{15}$
   (C) $\sqrt{72}$
   (D) $\sqrt{121}$

   A2.0

22. Solve the following inequality:
    $-3(4x+5) > (5x+6) + 13$

   (A) $x < -\frac{14}{17}$
   (B) $x > -2$
   (C) $x > \frac{20}{11}$
   (D) $x < -2$

   A4.0

23. Solve for a: $-2(-3-5) = 3 - a$

   (A) -13
   (B) 19
   (C) 13
   (D) -19

   A4.0

24. Find c: $\dfrac{c}{-2} > -6$

   (A) $c > -12$
   (B) $c < 12$
   (C) $c > 12$
   (D) $c < 3$

   A2.0

25. $(3x^2 - 5x + 6) - (x^2 + 4x - 7) =$

   (A) $4x^2 - x - 1$
   (B) $4x^2 - x + 13$
   (C) $2x^2 - 9x - 1$
   (D) $2x^2 - 9x + 13$

   A10.0

26. What graph shows a line that has a slope of -1 and a y-intercept of $(0, 1)$?

   (A)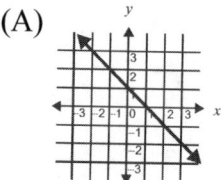

   (B)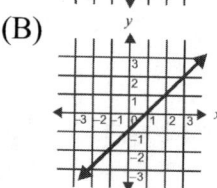

   (C)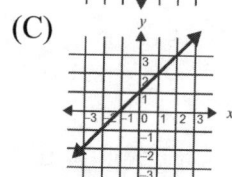

   (D)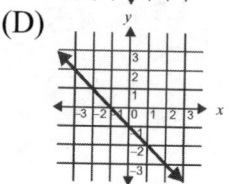

   A6.0

27. Solve the equation $(x-4)^2 = 25$.

   (A) $x = -4, -5$
   (B) $x = 9, -1$
   (C) $x = 5, -5$
   (D) $x = 4, -4$

   A14.0

28. Solve the equation $x^2 - 6x + 7 = 0$ by completing the square.

   (A) $x = 7, -1$
   (B) $x = \sqrt{3}, -\sqrt{3}$
   (C) $x = 3 - \sqrt{2}, \sqrt{2} + 3$
   (D) $x = 3 + 2i, 3 - 2i$

   A14.0

29. Solve the equation $x = \sqrt{4x - 3}$ by using the quadratic formula.

    (A) $x = 1, 3$
    (B) $x = \pm\sqrt{3}$
    (C) $x = -1, -3$
    (D) $x = 2\sqrt{3}, -2\sqrt{3}$

    A19.0

30. Fido gets 2 doggy treats every time he sits and 4 doggy treats when he rolls over on command. Throughout the week, he has sat 6 times as often as he has rolled over. In total, he has earned 80 doggy treats. How many times has Fido sat?

    (A) 4
    (B) 5
    (C) 24
    (D) 30

    A5.0

31. Solve the following quadratic equation by factoring: $3x^2 = 4x + 7$

    (A) $-\frac{1}{3}, \frac{7}{3}$
    (B) $-1, \frac{7}{3}$
    (C) $-1, -2$
    (D) $-\frac{7}{3}, \frac{3}{7}$

    A14.0

32. Multiply and simplify:

    $(3x + 2)(x - 4)$

    (A) $3x^2 - 10x - 8$
    (B) $3x^2 + 5x - 8$
    (C) $3x^2 + 5x - 6$
    (D) $8x^2 - 2$

    A10.0

33. Alexandria wants to locate the midpoint of a line segment with endpoints $(-3, -2)$ and $(6, -4)$. What are the coordinates of the midpoint?

    (A) $(1.5, -3)$
    (B) $(4.5, -3)$
    (C) $(4.5, -6)$
    (D) $(3, -6)$

    G17.0

34. Find the solution to $4m^2 = 9m + 9$ using the quadratic formula.

    (A) $\left\{-\frac{3}{2}, \frac{3}{2}\right\}$
    (B) $\left\{3, -\frac{3}{4}\right\}$
    (C) $\left\{-3, \frac{3}{4}\right\}$
    (D) $\left\{-1, \frac{1}{4}\right\}$

    A20.0

35. Solve the following quadratic equation by using the quadratic formula: $64x^2 = 25$

    (A) $\left\{-\frac{5}{8}, \frac{5}{8}\right\}$
    (B) $\left\{\frac{5}{8}, -\frac{8}{5}\right\}$
    (C) $\left\{\frac{8}{5}, -\frac{5}{8}\right\}$
    (D) $\left\{-\frac{8}{5}, -\frac{5}{8}\right\}$

    A20.0

36. Using the distance formula, what is the length of $\overline{AB}$?

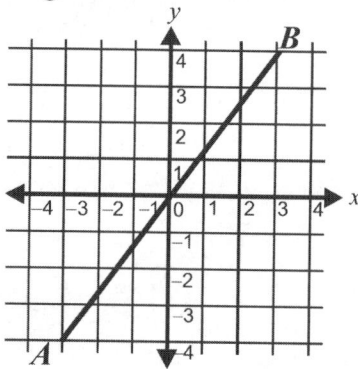

(A) $\sqrt{14}$
(B) 8
(C) 10
(D) 20

G17.0

37. The sum of two consecutive odd numbers is 96. What is the value of the larger number?

(A) 45
(B) 47
(C) 49
(D) 51

A5.0

38. Solve for y using the quadratic formula: $9y^2 - 64 = 0$

(A) $\left\{\dfrac{8}{9}, -\dfrac{8}{9}\right\}$
(B) $\left\{\dfrac{9}{8}, \dfrac{8}{9}\right\}$
(C) $\left\{\dfrac{3}{8}, -\dfrac{3}{8}\right\}$
(D) $\left\{\dfrac{8}{3}, -\dfrac{8}{3}\right\}$

A20.0

39. Factor $9x^2 + 15x - 14$.

(A) $(9x - 1)(x + 7)$
(B) $(3x + 2)(3x - 7)$
(C) $(3x - 2)(3x - 7)$
(D) $(3x - 2)(3x + 7)$

A11.0

40. Solve $a^2 + 5a - 14 = 0$ for a.

(A) The solutions are 2 and -7.
(B) The solutions are -2 and 7.
(C) The solutions are -2 and -7.
(D) The solutions are 2 and 7.

A19.0

41. Using the formula $A = \pi r^2$, Sally calculated the area of a round wading pool with a diameter of 10 feet to be 314 square feet. Maybelle believes this is not correct because

(A) $3 \times 5 = 15$
(B) $3 \times 10 = 30$
(C) $3 \times 5 \times 5 = 75$
(D) $3 \times 20 \times 20 = 1200$

G8.0

42. William draws a picture of his yard to help figure the area of the yard. He needs to know how much grass seed and fertilizer to buy. The shaded area represents his house. How many square feet of yard does he have?

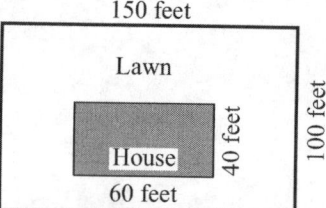

(A) 10,000 sq ft
(B) 12,600 sq ft
(C) 14,760 sq ft
(D) 15,000 sq ft

G10.0

43. Johnny is making a bird house. The front of the house is a regular pentagon.

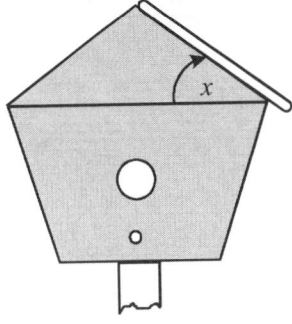

What is the measure of ∠x in degrees?

(A) 108°
(B) 36°
(C) 72°
(D) Cannot be determined.

G12.0

44. Which of the following is the graph of the equation $y = x - 3$?

(A)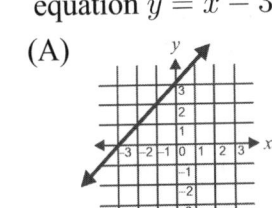

(B)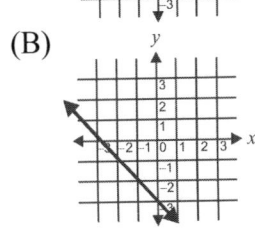

(C)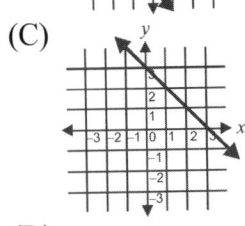

(D)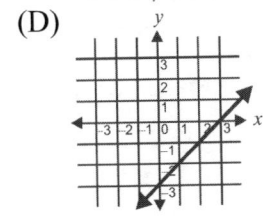

A6.0

45. Find the volume of the figure below.

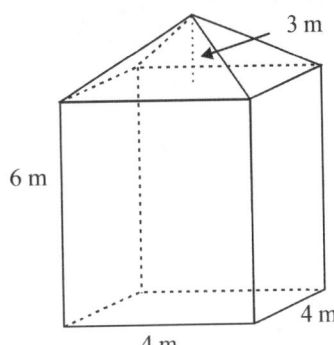

(A) 96 m³
(B) 99 m³
(C) 112 m³
(D) 288 m³

G8.0

46. If two triangles have all corresponding sides and all corresponding angles congruent, then they are congruent triangles. If two triangles are congruent, then they are similar triangles. Given these facts, which of the following statements is valid?

(A) Similar triangles have all sides and all angles congruent.
(B) If two triangles are similar, then they are congruent.
(C) If two triangles are not congruent, then they are not similar.
(D) If two triangles have all corresponding sides and angles congruent, then they are similar triangles.

A25.1

47. Which order of operations should be used to simplify the following expression?
$12 \div 2 + 4(7 - 5)$

(A) subtract, multiply, add, divide
(B) divide, add, subtract, multiply
(C) divide, add, multiply, subtract
(D) subtract, divide, multiply, add

A25.2

48. The figure below is a regular decagon.

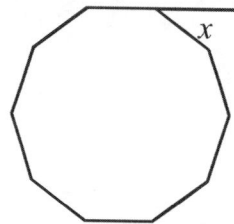

What is the measure of ∠x ?

(A) 10°
(B) 18°
(C) 25°
(D) 36°

G12.0

49. How many square feet is a 9-foot by 60-foot lawn?

(A) 69 square feet
(B) 138 square feet
(C) 270 square feet
(D) 540 square feet

G10.0

50. Triangle ABC, shown in the diagram below, is an isosceles triangle.

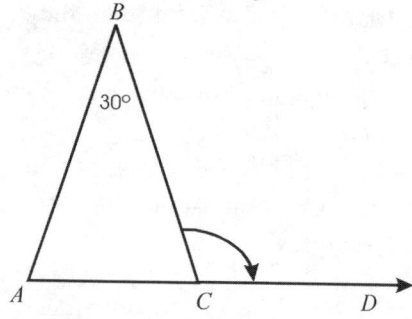

What is the measure, in degrees, of ∠BCD?

(A) 75°
(B) 105°
(C) 150°
(D) 165°

G12.0

51. What is the measure, in degrees, of ∠CBE?

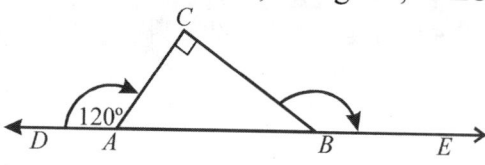

(A) 30°
(B) 60°
(C) 120°
(D) 150°

G12.0

52. $(-a^3 + 2a^2 - 8) + (-4a^3 + 5a - 2) =$

(A) $3a^3 + 2a^2 - 5a - 6$
(B) $-5a^3 + 7a^2 - 10$
(C) $3a^3 + 7a^2 - 6$
(D) $-5a^3 + 2a^2 + 5a - 10$

A10.0

53. Susanna is buying at least 150 sodas for the school dance party. She buys 2 cases with 24 sodas in each case. She also buys some six-packs of soda. Determine the number of six-packs Susanna buys using the following inequality.
$6x + 2(24) \geq 150$

(A) $x \geq 17$
(B) $x \geq 20$
(C) $x \geq 21$
(D) $x \geq 25$

A5.0

54.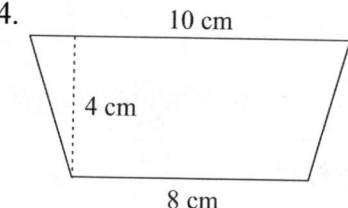

Find the area of the trapezoid above.

(A) 22 square centimeters
(B) 36 square centimeters
(C) 72 square centimeters
(D) 320 square centimeters

G10.0

55. Simplify: $\dfrac{4a^2 - 9}{10a^2 - 13a - 3}$

(A) $\dfrac{2a - 3}{5a - 1}$

(B) $\dfrac{(2a + 3)(2a - 3)}{(2a - 3)(5a + 1)}$

(C) $\dfrac{2a + 3}{5a + 1}$

(D) $\dfrac{2a + 3}{(2a - 3)(5a + 1)}$

A12.0

56. Divide $4x^2y^3 - 8xy^2 + 12xy$ by $4xy$.

(A) $x^2y^2 - 2xy^2 + 3xy$
(B) $4xy^2 - 8y + 12$
(C) $xy^2 - 8xy^2 + 12xy$
(D) $xy^2 - 2y + 3$

A13.0

57. Factor: $48xy^2 + 24xy^4 - 36xy^6$

(A) $12xy^2(4 + 2y^2 - 3y^4)$
(B) $24xy(2y + y^3 - 3y^5)$
(C) $12xy(4y + 2y^3 - 3y^5)$
(D) $12(4xy^2 + 2xy^4 - 3xy^6)$

A11.0

58. Using the triangle inequality theorem, which of the following measures of sides can form a triangle?

(A) 14, 15, 30
(B) 11, 21, 10
(C) 4, 6, 12
(D) 7, 8, 9

G6.0

59. Multiply and simplify the following.

$\dfrac{2xy}{5xy^2} \times \dfrac{3x^2}{4}$

(A) $\dfrac{6x^3y}{20xy}$

(B) $\dfrac{3x^2}{10y}$

(C) $\dfrac{6x^2}{20y}$

(D) $\dfrac{3x}{10y^2}$

A13.0

60. Add and simplify the following.

$\dfrac{2x^2}{5} + \dfrac{x^2}{4x}$

(A) $\dfrac{3x^2}{4x + 5}$

(B) $\dfrac{3x}{20}$

(C) $\dfrac{8x^3 + 5x^2}{20x}$

(D) $\dfrac{x(8x + 5)}{20}$

A13.0

61. Factor and simplify.

$\dfrac{x^2 - 5x + 6}{x - 2}$

(A) $x - 3$

(B) $\dfrac{(x - 2)(x - 3)}{x - 2}$

(C) $x - 2$

(D) Cannot be simplified

A12.0

62. What is the area of the parallelogram?

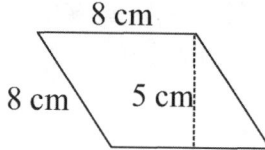

(A) 64 cm²
(B) 40 cm²
(C) 320 cm²
(D) 80 cm²

G10.0

63. Subtract and simplify the following.

$$\frac{2xy}{4} - \frac{5x^2}{y^3}$$

(A) $\dfrac{2xy^4 - 20x^2}{4y^3}$

(B) $\dfrac{x(1-10x)}{2}$

(C) $\dfrac{-5x^2 + 2xy}{4y^3}$

(D) $\dfrac{x(y^4 - 10x)}{2y^3}$

A13.0

64. What is the sum of the measure of the angles in the figure below?

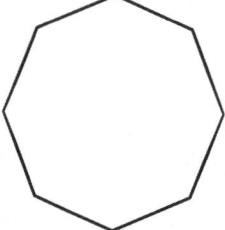

(A) 1440°
(B) 1260°
(C) 1080°
(D) 900°

G12.0

65. Find the equation of the line that passes through the points $(-2, 5)$ and $(0, 8)$.

(A) $y = \frac{3}{2}x + 8$

(B) $y = -\frac{13}{2}x + 8$

(C) $y = -\frac{13}{2}x - 8$

(D) $y = \frac{2}{3}x + 8$

A7.0

Evaluation Chart for the Diagnostic Mathematics Test

Directions: On the following chart on this page and the next, circle the question numbers that you answered incorrectly. Then turn to the appropriate topics (listed by chapters), read the explanations, and complete the exercises. Review the other chapters as needed. Finally, complete *California Integrated Mathematics 1* Practice Tests for further review.

		Questions	Pages
Chapter 1:	Exponents	1, 17, 47	12–26
Chapter 2:	Roots	21	27–33
Chapter 3:	Introduction to Algebra		34–45
Chapter 4:	Solving Multi-Step Equations and Inequalities	20, 22, 23, 24	46–57
Chapter 5:	Algebra Word Problems	5, 10, 14, 30, 37, 53	58–65
Chapter 6:	Polynomials	18, 25, 32, 52, 56	66–81
Chapter 7:	Factoring	12, 39, 55, 57, 59, 60, 61, 63	82–100
Chapter 8:	Solving Quadratic Equations	27, 28, 29, 31, 34, 35, 38, 40	101–110
Chapter 9:	Graphing and Writing Equations and Inequalities	4, 7, 9, 13, 15, 16, 19, 26, 33, 36, 44, 65	111–134
Chapter 10:	Mathematical Reasoning	46	135–140
Chapter 11:	Triangles	50, 51, 58	141–146
Chapter 12:	Plane Geometry	2, 11, 41, 42, 43, 48, 49, 54, 62, 64	147–161
Chapter 13:	Solid Geometry	3, 6, 8, 45	162–178

Chapter 1
Exponents

This chapter covers the following CA Integrated Math I standards:

Algebra	2.0
	25.2

1.1 Understanding Rational Numbers

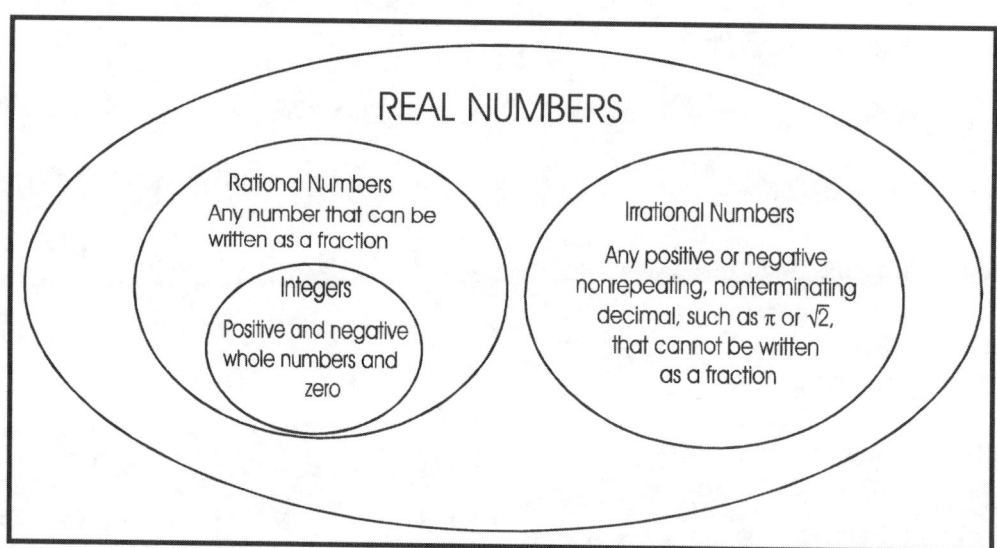

Real numbers include all positive and negative numbers and zero. Included in the set of real numbers are positive and negative fractions, decimals, and rational and irrational numbers.

A **rational number** is any number that can be expressed as $\frac{a}{b}$.

Whole numbers are rational numbers. $1 = \frac{1}{1}$ $2 = \frac{2}{1}$

Fractions are rational numbers. $\frac{1}{4}, \frac{2}{9}, \frac{1}{100}$

Mixed numbers are rational numbers. $1\frac{1}{2} = \frac{3}{2}$ $2\frac{1}{4} = \frac{9}{4}$

Decimals are rational numbers. $0.1 = \frac{1}{10}$ $0.014 = \frac{14}{1000}$

Negative numbers can be rational. $\frac{-5}{8}$

Use the diagram on the previous page and your calculator to answer the following questions.

1. Using your calculator, find the square root of 7. Does it repeat? Does it end? Is it a rational or an irrational number?
2. Find $\sqrt{25}$. Is it rational or irrational? is it an integer?
3. Is an integer an irrational number?
4. Is an integer a real number?
5. Is $\frac{1}{8}$ a real number? Is it rational or irrational?

Identify the following numbers as rational (R) or (I).

6. 5π
7. $\sqrt{8}$
8. $\frac{1}{3}$
9. -7.2
10. $-\frac{3}{4}$
11. $\frac{\sqrt{2}}{2}$
12. $9 + \pi$
13. 1.0004
14. $-\frac{4}{5}$
15. $1.1\overline{8}$
16. $\sqrt{81}$
17. $\frac{\pi}{4}$
18. $-\sqrt{36}$
19. $17\frac{1}{2}$
20. $-\frac{5}{3}$

1.2 Opposite Numbers

The **opposite** of a number is the negative of that number. For example, the opposite of 8 is -8. To find the opposite of a negative number, you also take the negative of that number. For example, to find the opposite of -2, take the negative of -2, $-(-2) = 2$. The opposite of -2 is 2. The opposite of a number is the same number, but different sign. The only number that does not have an opposite is 0.

If the number is positive, its opposite will be negative.
If the number is negative, its opposite will be positive.

Find the opposite of the numbers below.

1. 3
2. $\frac{1}{5}$
3. -19
4. 45
5. 1.01
6. $-\frac{2}{3}$
7. -2.4
8. $-3\frac{1}{2}$
9. 14
10. $\frac{5}{9}$
11. -62
12. 7
13. -1000
14. $-\frac{7}{4}$
15. 23
16. 1
17. 7
18. $\frac{6}{7}$
19. -2.3
20. $-5\frac{1}{3}$

Chapter 1 Exponents

1.3 Reciprocals

If the product of two numbers is 1, then the two numbers are **reciprocals** of each other. The numbers $\frac{a}{b}$ and $\frac{b}{a}$ are reciprocals because $\frac{a}{b} \times \frac{b}{a} = 1$. To find the reciprocal of a number, put one over that number. For example, the reciprocal of 4 is $\frac{1}{4}$. The reciprocal of a fraction is the fraction flipped upside down. For example, the reciprocal of $\frac{3}{4}$ is $\frac{4}{3}$. The sign of the number does not change when you find the reciprocal. The only number that does not have a reciprocal is 0. Finding the reciprocal of a number is also referred to as taking the multiplicative inverse of that number.

Find the reciprocal of the following numbers.

1. $\frac{5}{7}$
2. 1
3. -5
4. $\frac{6}{4}$
5. $-\frac{10}{3}$
6. 43
7. 15
8. $\frac{4}{5}$
9. $-\frac{1}{2}$
10. -2
11. 575
12. $\frac{1}{120}$
13. $\frac{4}{1}$
14. 7
15. -9
16. $\frac{2}{7}$
17. $-\frac{4}{9}$
18. 33
19. 12
20. $\frac{1}{6}$

1.4 Understanding Exponents

Sometimes it is necessary to multiply a number by itself one or more times. For example, a math problem may need to multiply 3×3 or $5 \times 5 \times 5 \times 5$. In these situations, mathematicians have come up with a shorter way of writing out this kind of multiplication. Instead of writing 3×3, you can write 3^2, or instead of writing $5 \times 5 \times 5 \times 5$, 5^4 means the same thing. The first number is the **base**. The small, raised number is called the **exponent** or **power**. The exponent tells how many times the base should be multiplied by itself.

Example 1: 6^3 ← exponent (or power)
← base
This means multiply by 6 three times: $6 \times 6 \times 6$

Example 2: $4^1 = 4 \qquad 10^1 = 10 \qquad 25^1 = 25 \qquad 4^0 = 1 \qquad 10^0 = 1 \qquad 25^0 = 1$

All rational numbers can have exponents.

Examples: $\left(\frac{1}{4}\right)^3 = \frac{1}{4} \times \frac{1}{4} \times \frac{1}{4} = \frac{1}{64}$

$\left(1\frac{1}{2}\right)^2 = \left(\frac{3}{2}\right)^2 = \frac{3}{2} \times \frac{3}{2} = \frac{9}{4}$

$0.2^3 = 0.2 \times 0.2 \times 0.2 = 0.008$

$\left(\frac{x}{y}\right)^2 = \frac{x}{y} \times \frac{x}{y} = \frac{x^2}{y^2}$

Rewrite the following problems using exponents.

Example 3: $2 \times 2 \times 2 = 2^3$

1. $7 \times 7 \times 7 \times 7$
2. 10×10
3. $12 \times 12 \times 12$
4. $4 \times 4 \times 4 \times 4$
5. $9 \times 9 \times 9$
6. 25×25

Use your calculator to determine what product each number with an exponent represents.

Example 4: $2^3 = 2 \times 2 \times 2 = 8$

7. 8^3
8. 12^2
9. 20^1
10. 5^4
11. 15^0
12. 16^2
13. 10^2
14. 3^5

Express each of the following numbers as a base with an exponent.

Example 5: $4 = 2 \times 2 = 2^2$

15. 9
16. 16
17. 27
18. 36
19. 8
20. 32
21. 1000
22. 125

1.5 Multiplying Exponents with the Same Base

To multiply two expressions with the same base, add the exponents together and keep the base the same.

Example 6: $2^3 \times 2^5 = (2 \times 2 \times 2)(2 \times 2 \times 2 \times 2 \times 2) = 2^8$

Example 7: $3a^2 \times 2a^3 = 6a^{2+3} = 6a^5$
Notice that only the "a" is raised to a power and not the 3 or the 2.

Simplify each of the expressions below.

1. $2^3 \times 2^5$
2. $x^5 \times x^3$
3. $2a^3 \times 3a^3$
4. $4^5 \times 4^3$
5. $2x^3 \times x^5$
6. $4b^3 \times 2b^4$
7. $10^5 \times 10^4$
8. $5^2 \times 5^4$
9. $3^3 \times 3^2$
10. $4x \times x^2$
11. $a^2 \times 3a^4$
12. $2^3 \times 2^4$

Chapter 1   Exponents

1.6   Multiplying Fractional Exponents with the Same Base

To multiply two expressions with the same base, add the exponents together and keep the base the same. Numbers with **fractional exponents** follow the same rules as numbers with whole numbers as the exponent.

Example 8: $(4)^{\frac{1}{2}} \times (4)^{\frac{3}{2}} = 4^{\frac{1}{2}+\frac{3}{2}} = 4^{\frac{4}{2}} = 4^2$

Example 9: $2x^{\frac{3}{4}} \times 5x^{\frac{1}{5}} = 10x^{\frac{3}{4}+\frac{1}{5}} = 10x^{\frac{15}{20}+\frac{4}{20}} = 10x^{\frac{19}{20}}$

Notice that only the "x" is raised to a power and not the 2 or the 5.

Simplify each of the expressions below.

1. $y^{\frac{1}{3}} \times y^{\frac{2}{3}}$
2. $(2)^{\frac{5}{9}} \times (2)^{\frac{1}{3}}$
3. $(10)^{\frac{4}{7}} \times (10)^{\frac{5}{7}}$
4. $b^{\frac{1}{2}} \times b^{\frac{1}{2}}$
5. $x^{\frac{1}{6}} \times x^{\frac{2}{3}}$
6. $(6)^{\frac{1}{12}} \times (6)^{\frac{3}{4}}$
7. $(3)^{\frac{5}{12}} \times (3)^{\frac{7}{2}}$
8. $(12)^{\frac{13}{25}} \times (12)^{\frac{1}{5}}$
9. $2a^{\frac{7}{9}} \times a^{\frac{4}{9}}$
10. $(8)^{\frac{2}{5}} \times (8)^{\frac{1}{10}}$
11. $4a^{\frac{6}{7}} \times 4a^{\frac{4}{5}}$
12. $3(x)^{\frac{4}{13}} \times 5(x)^{\frac{1}{2}}$

1.7   Multiplying Exponents Raised to an Exponent

If a power is raised to another power, multiply the exponents together and keep the base the same.

Example 10: $(2^3)^2 = (2^3)(2^3) = (2 \times 2 \times 2)(2 \times 2 \times 2) = 2^{3 \times 2} = 2^6$

Example 11: $(y^4)^3 = y^{4 \times 3} = y^{12}$

Simplify each of the expressions below.

1. $(5^3)^2$
2. $(x^5)^2$
3. $(6^2)^5$
4. $(3^4)^2$
5. $(3^2)^4$
6. $(y^2)^3$
7. $(3^3)^2$
8. $(9^2)^2$
9. $(5^3)^2$
10. $(a^4)^2$
11. $(3^4)^2$
12. $(x^3)^2$

1.8 Expressions Raised to a Power

If an expression is raised to a power, do not raise each term to the power, but rather consider the expression as a whole and raise it to the power.

Example 12: $(2+3)^2 = (2+3)(2+3) = (5)(5) = 25$

Example 13: $(2-5)^3 = (2-5)(2-5)(2-5) = (-3)(-3)(-3) = -27$

Simplify each of the expressions below.

1. $(2+6)^2$
2. $(1-2)^2$
3. $(3+2)^2$
4. $(4+5)^2$
5. $(5+1)^3$

6. $(2+3)^3$
7. $(4-6)^2$
8. $(2+1)^4$
9. $(25+15)^0$
10. $(4+1+7)^2$

11. $(5-9+2)^3$
12. $(2+1+4)^2$
13. $(-7+-7+6)^3$
14. $(2+2-5+4)^2$
15. $(20+45+40)^0$

1.9 Fractions Raised to a Power

A fraction can also be raised to a power.

Example 14: $\left(\dfrac{3}{4}\right)^3 = \dfrac{3^3}{4^3} = \dfrac{27}{64}$

Simplify the following fractions.

1. $\left(\dfrac{2}{3}\right)^2$
2. $\left(\dfrac{7}{8}\right)^3$
3. $\left(\dfrac{1}{2}\right)^2$

4. $\left(\dfrac{1}{4}\right)^2$
5. $\left(\dfrac{2}{3}\right)^3$
6. $\left(\dfrac{3}{4}\right)^2$

7. $\left(\dfrac{1}{2}\right)^3$
8. $\left(\dfrac{5}{7}\right)^2$
9. $\left(\dfrac{2}{3}\right)^4$

10. $\left(\dfrac{3}{10}\right)^2$
11. $\left(\dfrac{4}{5}\right)^2$
12. $\left(\dfrac{1}{10}\right)^4$

Chapter 1   Exponents

1.10   More Multiplying Exponents

If a product in parentheses is raised to a power, then each factor is raised to the power when parentheses are eliminated.

Example 15: $(2 \times 4)^2 = 2^2 \times 4^2 = 4 \times 16 = 64$

Example 16: $(3a)^3 = 3^3 \times a^3 = 27a^3$

Example 17: $(7b^5)^2 = 7^2 b^{10} = 49b^{10}$

Simplify each of the following.

1. $(2^3)^2$
2. $(7a^5)^2$
3. $(6b^2)^2$
4. $(3^2)^2$
5. $(3 \times 5)^2$
6. $(3x^4)^2$
7. $(6y^7)^2$
8. $(11w^3)^2$
9. $(3^3)^2$
10. $(3 \times 3)^2$
11. $(2a)^4$
12. $(2^2)^3$
13. $(3 \times 2)^3$
14. $(5^3)^2$
15. $(4r^7)^3$
16. $(2m^3)^2$
17. $(6 \times 4)^2$
18. $(9a^5)^2$
19. $(7b^5)^2$
20. $(9^2)^2$
21. 4×4^3
22. $(3a)^2$
23. $(2 \times 3)^3$
24. $(5p^4)^3$
25. $(4y^4)^2$
26. $(2b^3)^4$
27. $(5a^2)^2$
28. $(8a^3)^2$
29. $(2 \times 6)^2$
30. $(7^2)^2$

1.11   More Multiplying Fractional Exponents

If a product in parentheses is raised to a power, then each factor is raised to the power when parentheses are eliminated. This rule applies to fractional exponents as well.

Example 18: $(2x)^{\frac{2}{3}} = 2^{\frac{2}{3}} \times x^{\frac{2}{3}} = 2^{\frac{2}{3}} x^{\frac{2}{3}}$

Example 19: $(3a^2)^{\frac{1}{4}} = 3^{\frac{1}{4}} a^{\frac{2}{4}} = 3^{\frac{1}{4}} a^{\frac{1}{2}}$

Simplify each of the following.

1. $(5)^{\frac{1}{2}}$
2. $(7x)^{\frac{6}{7}}$
3. $(14x^2)^{\frac{2}{9}}$
4. $\left(x^{\frac{5}{7}}\right)^{\frac{1}{5}}$
5. $(x^3)^{\frac{1}{3}}$
6. $(4x^5)^{\frac{4}{25}}$
7. $(23y)^{\frac{1}{2}}$
8. $\left(3x^{\frac{1}{2}}\right)^{\frac{2}{5}}$
9. $(16)^{\frac{1}{2}}$
10. $(11x^2)^{\frac{5}{2}}$
11. $(8)^{\frac{6}{13}}$
12. $\left(4a^{\frac{7}{2}}\right)^{\frac{1}{2}}$
13. $(5t)^{\frac{9}{11}}$
14. $(z^9)^{\frac{1}{9}}$
15. $(17)^{\frac{3}{4}}$
16. $\left(9y^{\frac{1}{2}}\right)^{\frac{1}{3}}$
17. $(8)^{\frac{1}{4}}$
18. $(33x)^{\frac{4}{7}}$
19. $(x^4)^{\frac{3}{8}}$
20. $\left(12x^{\frac{7}{9}}\right)^{\frac{3}{7}}$
21. $(4w)^{\frac{1}{6}}$
22. $(z^2)^{\frac{2}{13}}$
23. $(15)^{\frac{1}{15}}$
24. $\left(5a^{\frac{2}{9}}\right)^{\frac{3}{10}}$

1.12 Negative Exponents

Expressions can also have negative exponents. Negative exponents do not indicate negative numbers. They indicate **reciprocals**. The **reciprocal** of a number is 1 divided by that number. For example, the reciprocal of 2 is $\frac{1}{2}$. (A number multiplied by its reciprocal is equal to 1.) If the negative exponent is in the bottom of a fraction (denominator), the reciprocal will put the expression on the top of the fraction (numerator) without a negative sign.

Example 20: $2^{-3} = \frac{1}{2^3} = \frac{1}{8}$

Example 21: $3a^{-5} = 3 \times \frac{1}{a^5} = \frac{3}{a^5}$ Notice that the 3 is not raised to the -5 power, only the a.

Example 22: $\frac{6}{5x^{-2}} = \frac{6x^2}{5}$ The 5 is not raised to the -2 power, only the x.

Rewrite using only positive exponents.

1. $5m^{-6}$
2. $\frac{5x^{-4}}{7}$
3. $14z^{-8}$
4. $\frac{1}{5s^{-4}}$
5. $14h^{-5}$
6. $\frac{h^{-3}}{5}$
7. $\frac{2y^{-3}}{4}$
8. x^{-4}
9. $-2y^{-2}$
10. $5y^{-5}$
11. $\frac{x^{-3}}{5}$
12. $10z^{-7}$
13. $7x^{-3}$
14. r^{-2}
15. $\frac{m^{-4}}{6}$

1.13 Multiplying with Negative Exponents

Multiplying with negative exponents follows the same rules as multiplying with positive exponents.

Example 23: $6^2 \cdot 6^{-3} = 6^{2+(-3)} = 6^{-1} = \frac{1}{6}$

Example 24: $(5a \times 2)^{-3} = (10a)^{-3} = \frac{1}{(10a)^3} = \frac{1}{1000a^3}$

Example 25: $(7a^2)^{-3} = 7^{-3}a^{-6} = \frac{1}{7^3 a^6}$

Simplify the following. Answers should <u>not</u> have any negative exponents.

1. $5^{-2} \cdot 5^5$
2. $(6^3 \cdot 6^{-2})^{-2}$
3. $10^{-4} \cdot 10^2$
4. $11^{-5} \cdot 11^7$
5. $4^7 \cdot 4^{-10}$
6. $20^8 \cdot 20^{-6}$
7. $5^{-8} \cdot 5^4$
8. $(2^{-2} \cdot 2^3)^{-4}$
9. $7^{-2} \cdot 7^{-1}$
10. $(3x^4)^{-3}$
11. $12^{-10} \cdot 12^8$
12. $(10^8 \cdot 10^{-10})^2$
13. $3^{-2} \cdot 2^{-2}$
14. $(8x^5)^{-4}$
15. $(6b^3)^{-6}$
16. $(9y)^{-2}$

Chapter 1 Exponents

1.14 Dividing with Exponents

Exponents that have the same base can also be divided.

Example 26: $\dfrac{3^5}{3^3}$ This problem means $3^5 \div 3^3$. Let us look at 2 ways to solve this problem.

Solution 1: $\dfrac{3^5}{3^3} = \dfrac{3 \cdot 3 \cdot 3 \cdot 3 \cdot 3}{3 \cdot 3 \cdot 3} = 3 \cdot 3 = 9$ First, rewrite the fraction with the exponents in expanded form, and then multiply.

Solution 2: $\dfrac{3^5}{3^3} = 3^{5-3} = 3^2 = 9$ A quick way to simplify this same problem is to subtract the exponents. **When dividing exponents with the same base, subtract the exponents.**

Example 27: $\dfrac{(4x)^{-3}}{2x^4}$

Step 1: $(4x)^{-3} = \dfrac{1}{(4x)^3} = \dfrac{1}{4^3 x^3}$ Remove the parentheses from the top of the fraction.

Step 2: $\dfrac{1}{4^3 x^3 \cdot 2x^4} = \dfrac{1}{128 x^7}$ The bottom of the fraction remains the same, so put the two together and simplify.

Simplify the problems below. You may be able to cancel. Be sure to follow order of operations. Remove parentheses before canceling.

1. $\dfrac{5^5}{5^3}$

2. $\dfrac{x^2}{x^3}$

3. $\dfrac{(10^2)^4}{10^5}$

4. $\dfrac{3^5}{3^2}$

5. $\dfrac{8^{10}}{8^8}$

6. $\dfrac{5^2}{5}$

7. $\dfrac{(7^2)^3}{7^5}$

8. $\dfrac{(x^3)^4}{x^6}$

9. $\dfrac{4^3}{4^2}$

10. $\dfrac{2}{(2^2)^2}$

11. $\dfrac{(3x)^{-2}}{9x^5}$

12. $\dfrac{(11^4)^4}{(11^7)^2}$

13. $\dfrac{x^3}{(x^2)^3}$

14. $\dfrac{2^2}{2^7}$

15. $\dfrac{6^2}{6}$

16. $\dfrac{9^{11}}{9^9}$

17. $\dfrac{(15)^5}{15^6}$

18. $\dfrac{(x^3)^{-2}}{(x^2)^5}$

19. $\dfrac{12^{-4}}{12^{-2}}$

20. $\dfrac{6^{12}}{6^9}$

21. $\dfrac{8^8}{8^{10}}$

22. $\dfrac{3(x^{-3})^{-2}}{3x^7}$

23. $\dfrac{7^3}{7^5}$

24. $\dfrac{10^3}{10^{-1}}$

1.15 Dividing with Fractional Exponents

Fractional exponents that have the same base can also be divided.

Example 28: $\dfrac{5^{\frac{3}{4}}}{5^{\frac{1}{2}}}$ This problem means $5^{\frac{3}{4}} \div 5^{\frac{1}{2}}$.

Solution: $\dfrac{5^{\frac{3}{4}}}{5^{\frac{1}{2}}} = 5^{\frac{3}{4}-\frac{1}{2}} = 5^{\frac{3}{4}-\frac{2}{4}} = 5^{\frac{1}{4}}$

A quick way to simplify this problem is to subtract the exponents. **When dividing exponents with the same base, subtract the exponents.**

Simplify the problems below.

1. $\dfrac{x^{\frac{7}{9}}}{x^{\frac{1}{9}}}$

2. $\dfrac{2^{\frac{3}{4}}}{2^{\frac{3}{5}}}$

3. $\dfrac{12^{\frac{6}{7}}}{12^{\frac{1}{2}}}$

4. $\dfrac{a^{\frac{1}{3}}}{a^{\frac{1}{6}}}$

5. $\dfrac{6^{\frac{3}{5}}}{6^{\frac{1}{7}}}$

6. $\dfrac{x^{\frac{13}{25}}}{x^{\frac{2}{5}}}$

7. $\dfrac{y^{\frac{8}{9}}}{y^{\frac{2}{3}}}$

8. $\dfrac{8^{\frac{3}{4}}}{8^{\frac{1}{5}}}$

9. $\dfrac{2^{\frac{14}{15}}}{2^{\frac{4}{15}}}$

10. $\dfrac{7^{\frac{6}{7}}}{7^{\frac{2}{7}}}$

11. $\dfrac{b^{\frac{1}{2}}}{b^{\frac{1}{6}}}$

12. $\dfrac{5^{\frac{9}{10}}}{5^{\frac{2}{5}}}$

13. $\dfrac{4^{\frac{9}{17}}}{4^{\frac{1}{2}}}$

14. $\dfrac{9^{\frac{13}{15}}}{9^{\frac{2}{3}}}$

15. $\dfrac{17^{\frac{2}{5}}}{17^{\frac{1}{7}}}$

16. $\dfrac{3^{\frac{2}{3}}}{3^{\frac{1}{12}}}$

17. $\dfrac{x^{\frac{3}{4}}}{x^{\frac{4}{9}}}$

18. $\dfrac{2^{\frac{3}{2}}}{2^{\frac{9}{10}}}$

19. $\dfrac{16^{\frac{5}{3}}}{16^{\frac{7}{8}}}$

20. $\dfrac{10^{\frac{3}{24}}}{10^{\frac{1}{24}}}$

Chapter 1   Exponents

1.16   Order of Operations

In long math problems with $+$, $-$, $\times$, $\div$, $()$, and exponents in them, you have to know what to do first. Without following the same rules, you could get different answers. If you will memorize the silly sentence, Please Excuse My Dear Aunt Sally, you can memorize the order you must follow.

Please   "P" stands for parentheses. You must get rid of parentheses first.
Examples: $3(1+4) = 3(5) = 15$
$6(10-6) = 6(4) = 24$

Excuse   "E" stands for exponents. You must eliminate exponents next.
Example: $4^2 = 4 \times 4 = 16$

My Dear   "M" stands for multiply. "D" stands for divide. Start on the left of the equation and perform all multiplications and divisions in the order in which they appear.

Aunt Sally   "A" stands for add. "S" stands for subtract. Start on the left and perform all additions and subtractions in the order they appear.

Example 29: $12 \div 2(6-3) + 3^2 - 1$

Please	Eliminate **parentheses**. $6 - 3 = 3$ so now we have	$12 \div 2 \times 3 + 3^2 - 1$
Excuse	Eliminate **exponents**. $3^2 = 9$ so now we have	$12 \div 2 \times 3 + 9 - 1$
My Dear	**Multiply** and **divide** next in order from left to right.	$12 \div 2 = 6$ then $6 \times 3 = 18$
Aunt Sally	Last, we **add** and **subtract** in order from left to right.	$18 + 9 - 1 = 26$

Simplify the following problems.

1. $6 + 9 \times 2 - 4$
2. $3(4+2) - 6^2$
3. $3(6-3) - 2^3$
4. $49 \div 7 - 3 \times 3$
5. $10 \times 4 - (7-2)$
6. $2 \times 3 \div 6 \times 4$
7. $4^3 \div 8(4+2)$
8. $7 + 8(14-6) \div 4$
9. $(2 + 8 - 12) \times 4$
10. $4(8-13) \times 4$
11. $8 + 4^2 \times 2 - 6$
12. $3^2(4+6) + 3$
13. $(12-6) + 27 \div 3^2$
14. $82^0 - 1 + 4 \div 2^2$
15. $1 - (2-3) + 8$
16. $-4\{18 - (4 + 2 \times 6)\}$
17. $18 \div (6+3) - 12$
18. $10^2 + 3^3 - 2 \times 3$
19. $4^2 + (7+2) \div 3$
20. $7 \times 4 - 9 \div 3$

1.16 Order of Operations

When a problem has a fraction bar, simplify the top of the fraction (numerator) and the bottom of the fraction (denominator) separately using the rules for order of operations. You treat the top and bottom as if they were separate problems. Then reduce the fraction to lowest terms.

Example 30: $\dfrac{2(4-3)-6}{5^2+3(2+1)}$

Please	Eliminate **parentheses**. $(4-3)=1$ and $(2+1)=3$	$\dfrac{2 \times 1 - 6}{5^2 + 3 \times 3}$
Excuse	Eliminate **exponents**. $5^2 = 25$	$\dfrac{2 \times 1 - 6}{25 + 3 \times 3}$
My Dear	**Multiply** and **divide** in the numerator and denominator separately. $3 \times 3 = 9$ and $2 \times 1 = 2$	$\dfrac{2-6}{25+9}$
Aunt Sally	**Add** and **subtract** in the numerator and denominator separately. $2-6=-4$ and $25+9=34$	$\dfrac{-4}{34}$

Now reduce the fraction to lowest terms. $\dfrac{-4}{34} = \dfrac{-2}{17}$

Simplify the following problems.

1. $\dfrac{2^2+4}{5+3(8+1)}$

2. $\dfrac{8^2-(4+11)}{4^2-3^2}$

3. $\dfrac{5-2(4-3)}{2(1-8)}$

4. $\dfrac{10+(2-4)}{4(2+6)-2^2}$

5. $\dfrac{3^3-8(1+2)}{-10-(3+8)}$

6. $\dfrac{(9-3)+3^2}{-5-2(4+1)}$

7. $\dfrac{16-3(10-6)}{(13+15)-5^2}$

8. $\dfrac{(2-5)-11}{12-2(3+1)}$

9. $\dfrac{7+(8-16)}{6^2-5^2}$

10. $\dfrac{16-(12-3)}{8(2+3)-5}$

11. $\dfrac{-3(9-7)}{7+9-2^3}$

12. $\dfrac{4-(2+7)}{13+(6-9)}$

13. $\dfrac{5(3-8)-2^2}{7-3(6+1)}$

14. $\dfrac{3(3-8)+5}{8^2-(5+9)}$

15. $\dfrac{6^2-4(7+3)}{8+(9-3)}$

Chapter 1   Exponents

Chapter 1 Review

Simplify the following expressions. Simplify the answers. Make all exponents positive.

1. $5^2 \times 5^3$

2. $(4^4)^5$

3. $(4y^3)^3$

4. $6x^{-3}$

5. $(3a^2)^{-2}$

6. $(b^3)^{-4}$

7. $\dfrac{4^6}{4^4}$

8. $\left(\dfrac{3}{5}\right)^2$

9. $x^3 \cdot x^{-7}$

10. $(2x)^{-4}$

11. $3^3 \times 3^2$

12. $(2^4)^2$

13. $5^7 \times 5^{-4}$

14. $\dfrac{(3a^2)^3}{a^3}$

15. $(4^2)^{-2}$

16. $(5^{-9} \times 5^7)^{-2}$

17. $\dfrac{(2^3)^2}{2^4}$

18. $\dfrac{y^{-2}}{3y^4}$

19. $(6x)^{-3}$

20. $\dfrac{x^{\frac{6}{5}}}{x^{\frac{7}{15}}}$

21. $(4d^5)^{-3}$

22. $(5)^{\frac{3}{4}} \times (5)^{\frac{1}{3}}$

23. $\left(3^{\frac{1}{4}}\right)^{\frac{3}{5}}$

24. $(6)^{\frac{1}{12}} \times (6)^{\frac{1}{4}}$

25. $(3a^2)^{\frac{3}{2}}$

26. $\dfrac{8^{\frac{8}{9}}}{8^{\frac{2}{9}}}$

27. $(19)^{\frac{6}{13}}$

28. $(z)^{\frac{5}{7}} \times (z)^{\frac{5}{14}}$

29. $\left(5^{\frac{9}{13}}\right)^{\frac{1}{3}}$

30. $\dfrac{10^{\frac{3}{4}}}{10^{\frac{1}{12}}}$

Write using exponents.

31. $3 \times 3 \times 3 \times 3$

32. $6 \times 6 \times 6 \times 6 \times 6 \times 6$

33. $11 \times 11 \times 11$

34. $2 \times 2 \times 2 \times 2 \times 2 \times 2 \times 2 \times 2$

Simplify the following problems using the correct order of operations.

35. $10 \div (-1 - 4) + 2$

36. $5 + (2)(4 - 1) \div 3$

37. $5 - 5^2 + (2 - 4)$

38. $(8 - 10) \times (5 + 3) - 10$

39. $\dfrac{10 + 5^2 - 3}{2^2 + 2(5 - 3)}$

40. $1 - (9 - 1) \div 2$

41. $\dfrac{5(3 - 6) + 3^2}{4(2 + 1) - 6}$

42. $-4(6 + 4) \div (-2) + 1$

43. $12 \div (7 - 4) - 2$

44. $1 + 4^2 \div (3 + 1)$

Chapter 1 Test

1. Simplify the expression shown below:

   $$\frac{8x^4}{2x^2}$$

   (A) $2x^4$

   (B) $4x^2$

   (C) $\frac{1}{4x^2}$

   (D) $\frac{4x^2}{x}$

2. Simplify the following:

   $5 \cdot x^4 \cdot y^5 \cdot z^{-3}$

   (A) $\frac{5x^4y^5}{z^3}$

   (B) $(5xyz)^6$

   (C) $\frac{625x^4y^5}{z^3}$

   (D) $x^{20}y^{25}z^{-15}$

3. What is the solution to $2(5-2)^2 - 15 \div 5$?

   (A) $-\frac{3}{5}$

   (B) $\frac{3}{5}$

   (C) 15

   (D) $4\frac{1}{5}$

4. Simplify: $(6)^{\frac{5}{6}} \times (6)^{\frac{1}{3}}$

   (A) $(6)^{\frac{7}{6}}$

   (B) $(6)^{\frac{5}{18}}$

   (C) $(6)^{\frac{6}{9}}$

   (D) $(36)^{\frac{7}{6}}$

5. Simplify: $3^2 + 4 \times 18 \div 9$

   (A) 4

   (B) 14

   (C) 17

   (D) 26

6. $4 \times 18 - 9 \div 3^2 =$

   (A) 4

   (B) 7

   (C) 68

   (D) 71

7. Simplify: $\frac{(4x)^{-3}}{6x^4}$

   (A) $\frac{2}{3x^{12}}$

   (B) $\frac{2}{3x^7}$

   (C) $\frac{1}{384x^7}$

   (D) $-\frac{32}{3x}$

8. Simplify: $(6x^4)^{-2}$

   (A) $\frac{1}{36x^8}$

   (B) $-36x^8$

   (C) $-12x^{-8}$

   (D) $6x^{-8}$

9. $(3a^2)^{\frac{3}{4}} =$

   (A) $3^{\frac{3}{4}} a^{\frac{5}{4}}$

   (B) $3^{\frac{3}{4}} a^{\frac{3}{2}}$

   (C) $3a^{\frac{11}{4}}$

   (D) $3a^{\frac{3}{2}}$

Chapter 1   Exponents

10. $x^2 \cdot x^4 =$

(A) x^8
(B) $8x$
(C) x^6
(D) $6x$

11. Write using exponents: $4a \times 4a \times 4a$

(A) $(4a)^3$
(B) $64a^3$
(C) $3(4a)$
(D) $(4+a)^3$

12. Simplify: $(3^4)^2$

(A) 3^8
(B) 3^6
(C) 12^2
(D) 7^2

13. $(5y)^{-4} =$

(A) $\dfrac{1}{(5y)^4}$
(B) $\dfrac{1}{5y^4}$
(C) $-20 - 4y$
(D) $\dfrac{5}{y^4}$

14. $\dfrac{12^{\frac{7}{8}}}{12^{\frac{1}{4}}} =$

(A) $1^{\frac{5}{8}}$
(B) $12^{\frac{7}{2}}$
(C) $12^{\frac{5}{8}}$
(D) $12^{\frac{9}{8}}$

15. $3^3 x^2 \cdot 4x^5 =$

(A) $108x^{10}$
(B) $108x^5$
(C) $108x^7$
(D) $12x^{10}$

16. $3^2 \cdot 3^{-3} =$

(A) 9^{-6}
(B) $\dfrac{1}{3}$
(C) $\dfrac{1}{9}$
(D) 3

17. $8y^{-2} =$

(A) $\dfrac{1}{8y^2}$
(B) $6y$
(C) $-16y$
(D) $\dfrac{8}{y^2}$

18. $(4+2)^3 =$

(A) 18
(B) 72
(C) 216
(D) 54

Chapter 2
Roots

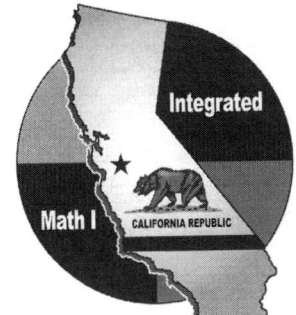

This chapter covers the following CA Integrated Math I standards:

| Algebra | 2.0 |

2.1 Square Root

Just as working with exponents is related to multiplication, finding square roots is related to division. In fact, the sign for finding the square root of a number looks similar to a division sign. The best way to learn about square roots is to look at examples.

Examples: This is a square root problem: $\sqrt{64}$
It is asking, "What is the square root of 64?"
It means, "What number multiplied by itself equals 64?"
The answer is 8. $8 \times 8 = 64$.

Find the square roots of the following numbers.

1. $\sqrt{49}$
2. $\sqrt{81}$
3. $\sqrt{25}$
4. $\sqrt{16}$
5. $\sqrt{121}$
6. $\sqrt{625}$
7. $\sqrt{100}$
8. $\sqrt{289}$
9. $\sqrt{196}$
10. $\sqrt{36}$
11. $\sqrt{4}$
12. $\sqrt{900}$
13. $\sqrt{64}$
14. $\sqrt{9}$
15. $\sqrt{144}$

2.2 Simplifying Square Roots

Square roots can sometimes be simplified even if the number under the square root is not a perfect square. One of the rules of roots is that if a and b are two positive real numbers, then it is always true that $\sqrt{a \cdot b} = \sqrt{a} \cdot \sqrt{b}$. You can use this rule to simplify square roots.

Example 1: $\sqrt{100} = \sqrt{4 \cdot 25} = \sqrt{4} \cdot \sqrt{25} = 2 \cdot 5 = 10$

Example 2: $\sqrt{200} = \sqrt{100 \cdot 2} = 10\sqrt{2}$ ⟵ Means 10 multiplied by the square root of 2

Example 3: $\sqrt{160} = \sqrt{10 \cdot 16} = 4\sqrt{10}$

Simplify.

1. $\sqrt{98}$
2. $\sqrt{600}$
3. $\sqrt{50}$
4. $\sqrt{27}$
5. $\sqrt{8}$
6. $\sqrt{63}$
7. $\sqrt{48}$
8. $\sqrt{75}$
9. $\sqrt{54}$
10. $\sqrt{40}$
11. $\sqrt{72}$
12. $\sqrt{80}$
13. $\sqrt{90}$
14. $\sqrt{175}$
15. $\sqrt{18}$
16. $\sqrt{20}$

2.3 Adding and Subtracting Roots

You can add and subtract terms with square roots only if the number under the square root sign is the same.

Example 4: $2\sqrt{2} + 3\sqrt{2} = 5\sqrt{2}$

Example 5: $12\sqrt{7} - 3\sqrt{7} - 9\sqrt{7}$

Or, look at the following examples where you can simplify the square roots and then add or subtract.

Example 6: $2\sqrt{25} + \sqrt{36}$

    Step 1: Simplify. You know that $\sqrt{25} = 5$, and $\sqrt{36} = 6$ so the problem simplifies to $2(5) + 6$

    Step 2: Solve: $2(5) + 6 = 10 + 6 = 16$

Example 7: $2\sqrt{72} - 3\sqrt{2}$

    Step 1: Simplify what you know. $\sqrt{72} = \sqrt{36 \cdot 2} = 6\sqrt{2}$

    Step 2: Substitute $6\sqrt{2}$ for $\sqrt{72}$ simplify.
$2(6)\sqrt{2} - 3\sqrt{2} = 12\sqrt{2} - 3\sqrt{2} = 9\sqrt{2}$

Simplify the following addition and subtraction problems.

1. $3\sqrt{5} + 9\sqrt{5}$
2. $3\sqrt{25} + 4\sqrt{16}$
3. $4\sqrt{8} + 2\sqrt{2}$
4. $3\sqrt{32} - 2\sqrt{2}$
5. $\sqrt{25} - \sqrt{49}$
6. $2\sqrt{5} + 4\sqrt{20}$
7. $5\sqrt{8} - 3\sqrt{72}$
8. $\sqrt{27} + 3\sqrt{27}$

9. $3\sqrt{20} - 4\sqrt{45}$
10. $4\sqrt{45} - \sqrt{75}$
11. $2\sqrt{28} + 2\sqrt{7}$
12. $\sqrt{64} + \sqrt{81}$
13. $5\sqrt{54} - 2\sqrt{24}$
14. $\sqrt{32} + 2\sqrt{50}$
15. $2\sqrt{7} + 4\sqrt{63}$
16. $8\sqrt{2} + \sqrt{8}$

17. $2\sqrt{8} - 4\sqrt{32}$
18. $\sqrt{36} + \sqrt{100}$
19. $\sqrt{9} + \sqrt{25}$
20. $\sqrt{64} - \sqrt{36}$
21. $\sqrt{75} + \sqrt{108}$
22. $\sqrt{81} + \sqrt{100}$
23. $\sqrt{192} - \sqrt{75}$
24. $3\sqrt{5} + \sqrt{245}$

2.4 Multiplying Roots

You can also multiply square roots. To multiply square roots, you just multiply the numbers under the square root sign and then simplify. Look at the examples below.

Example 8: $\sqrt{2} \times \sqrt{6}$

    Step 1: $\sqrt{2} \times \sqrt{6} = \sqrt{2 \times 6} = \sqrt{12}$ Multiply the numbers under the square root sign.

    Step 2: $\sqrt{12} = \sqrt{4 \times 3} = 2\sqrt{3}$ Simplify

Example 9: $3\sqrt{3} \times 5\sqrt{6}$

    Step 1: $(3 \times 5)\sqrt{3 \times 6} = 15\sqrt{18}$ Multiply the numbers in front of the square root, and multiply the numbers under the square root sign.

    Step 2: $15\sqrt{18} = 15\sqrt{2 \times 9}$
                $15 \times 3\sqrt{2} = 45\sqrt{2}$ Simplify.

Example 10: $\sqrt{14} \times \sqrt{42}$ For this more complicated multiplication problem, use the rule of roots that you learned on page 27, $\sqrt{a \cdot b} = \sqrt{a} \cdot \sqrt{b}$.

    Step 1: $\sqrt{14} = \sqrt{7} \times \sqrt{2}$ and
            $\sqrt{42} = \sqrt{2} \times \sqrt{3} \times \sqrt{7}$ Instead of multiplying 14 by 42, divide these numbers into their roots.

            $\sqrt{14} \times \sqrt{42} = \sqrt{7} \times \sqrt{2} \times \sqrt{2} \times \sqrt{3} \times \sqrt{7}$

    Step 2: Since you know that $\sqrt{7} \times \sqrt{7} = 7$ and $\sqrt{2} \times \sqrt{2} = 2$, the problem simplifies to $(7 \times 2)\sqrt{3} = 14\sqrt{3}$

Simplify the following multiplication problems.

1. $\sqrt{5} \times \sqrt{7}$
2. $\sqrt{32} \times \sqrt{2}$
3. $\sqrt{10} \times \sqrt{14}$
4. $2\sqrt{3} \times 3\sqrt{6}$
5. $4\sqrt{2} \times 2\sqrt{10}$

6. $\sqrt{5} \times 3\sqrt{15}$
7. $\sqrt{45} \times \sqrt{27}$
8. $5\sqrt{21} \times \sqrt{7}$
9. $\sqrt{42} \times \sqrt{21}$
10. $4\sqrt{3} \times 2\sqrt{12}$

11. $\sqrt{56} \times \sqrt{24}$
12. $\sqrt{11} \times 2\sqrt{33}$
13. $\sqrt{13} \times \sqrt{26}$
14. $2\sqrt{2} \times 5\sqrt{5}$
15. $\sqrt{6} \times \sqrt{12}$

2.5 Dividing Roots

When dividing a number or a square root by another square root, you cannot leave the square root sign in the denominator (the bottom number) of a fraction. You must simplify the problem so that the square root is not in the denominator. This is also called rationalizing the denominator. Look at the examples below.

Example 11: $\dfrac{\sqrt{2}}{\sqrt{5}}$

Step 1: $\dfrac{\sqrt{2}}{\sqrt{5}} \times \dfrac{\sqrt{5}}{\sqrt{5}}$ ← The fraction $\dfrac{\sqrt{5}}{\sqrt{5}}$ is equal to 1, and multiplying by 1 does not change the value of a number.

Step 2: $\dfrac{\sqrt{2 \times 5}}{5} = \dfrac{\sqrt{10}}{5}$ Multiply and simplify. Since $\sqrt{5} \times \sqrt{5}$ equals 5, you no longer have a square root in the denominator.

Example 12: $\dfrac{6\sqrt{2}}{2\sqrt{10}}$ In this problem, the numbers outside of the square root will also simplify.

Step 1: $\dfrac{6}{2} = 3$ so you have $\dfrac{3\sqrt{2}}{\sqrt{10}}$

Step 2: $\dfrac{3\sqrt{2}}{\sqrt{10}} \times \dfrac{\sqrt{10}}{\sqrt{10}} = \dfrac{3\sqrt{2 \times 10}}{10} = \dfrac{3\sqrt{20}}{10}$

Step 3: $\dfrac{3\sqrt{20}}{10}$ will further simplify because $\sqrt{20} = 2\sqrt{5}$, so you then have $\dfrac{3 \times 2\sqrt{5}}{10}$ which reduces to $\dfrac{3\sqrt{5}}{5}$.

Simplify the following division problems.

1. $\dfrac{9\sqrt{3}}{\sqrt{5}}$
2. $\dfrac{16}{\sqrt{8}}$
3. $\dfrac{24\sqrt{10}}{12\sqrt{3}}$
4. $\dfrac{\sqrt{121}}{\sqrt{6}}$

5. $\dfrac{\sqrt{40}}{\sqrt{90}}$
6. $\dfrac{33\sqrt{15}}{11\sqrt{2}}$
7. $\dfrac{\sqrt{32}}{\sqrt{12}}$
8. $\dfrac{\sqrt{11}}{\sqrt{5}}$

9. $\dfrac{\sqrt{2}}{\sqrt{6}}$
10. $\dfrac{2\sqrt{7}}{\sqrt{14}}$
11. $\dfrac{5\sqrt{2}}{4\sqrt{8}}$
12. $\dfrac{4\sqrt{21}}{7\sqrt{7}}$

13. $\dfrac{9\sqrt{22}}{2\sqrt{2}}$
14. $\dfrac{\sqrt{35}}{2\sqrt{14}}$
15. $\dfrac{\sqrt{40}}{\sqrt{15}}$
16. $\dfrac{\sqrt{3}}{\sqrt{12}}$

2.6 Cube Roots

Cube roots look like square roots, except that there is a "3" raised in front of the root sign:

Square root of 64: $\sqrt{64}$
Cube root of 64: $\sqrt[3]{64}$

In fact, they function very much like square roots, with one important difference. Recall asking, "What is the square root of 64?" means:
"What number multiplied by itself equals 64?"

Well, asking "What is the cube root of 64?" means:
"What number multiplied 3 times ('cubed') by itself equals 64?"
The answer is 4. $4 \times 4 \times 4 = 64$.

Find the cube root of the following numbers.

Examples: $\sqrt[3]{27}$ $3 \times 3 \times 3 = 27$ so $\sqrt[3]{27} = 3$
$\sqrt[3]{1000}$ $10 \times 10 \times 10 = 1000$ so $\sqrt[3]{1000} = 10$

Find the cube roots of the following numbers.

1. $\sqrt[3]{1}$
2. $\sqrt[3]{8}$
3. $\sqrt[3]{64}$
4. $\sqrt[3]{125}$
5. $\sqrt[3]{27}$
6. $\sqrt[3]{\frac{64}{27}}$
7. $\sqrt[3]{1000}$
8. $\sqrt[3]{\frac{125}{1000}}$

2.7 Fractional Exponents

Roots can be written as powers.

Example 13:
$\sqrt{2} = 2^{\frac{1}{2}}$
$\sqrt[3]{10} = 10^{\frac{1}{3}}$
$\sqrt[3]{3^2} = 3^{\frac{2}{3}}$
$\sqrt[5]{4^2} = 4^{\frac{2}{5}}$

Rewrite each of the following with rational (fractional) exponents.

1. $\sqrt{5}$
2. $\sqrt[3]{19}$
3. $\sqrt[4]{2^3}$
4. $\sqrt[5]{9^2}$
5. $\sqrt[3]{20}$
6. $\sqrt[4]{11^3}$
7. $\sqrt{5^3}$
8. $\sqrt[4]{7^3}$
9. $\sqrt[5]{2^3}$
10. $\sqrt{50}$
11. $\sqrt[3]{8^2}$
12. $\sqrt[5]{9^3}$

Chapter 2   Roots

Chapter 2 Review

Simplify the following square root expressions.

1. $\sqrt{50}$
2. $\sqrt{44}$
3. $\sqrt{12}$
4. $\sqrt{18}$
5. $\sqrt{8}$
6. $\sqrt{48}$
7. $\sqrt{75}$
8. $\sqrt{200}$
9. $\sqrt{32}$
10. $\sqrt{20}$
11. $\sqrt{63}$
12. $\sqrt{80}$

Simplify the following square root problems.

13. $5\sqrt{27} + 7\sqrt{3}$
14. $\sqrt{40} - \sqrt{10}$
15. $\sqrt{64} + \sqrt{81}$
16. $8\sqrt{50} - 3\sqrt{32}$
17. $14\sqrt{5} + 8\sqrt{80}$
18. $\sqrt{63} \times \sqrt{28}$
19. $\dfrac{\sqrt{56}}{\sqrt{35}}$
20. $\sqrt{8} \times \sqrt{50}$
21. $\dfrac{\sqrt{20}}{\sqrt{45}}$
22. $5\sqrt{40} \times 3\sqrt{20}$
23. $2\sqrt{48} - \sqrt{12}$
24. $\dfrac{2\sqrt{5}}{\sqrt{30}}$
25. $\dfrac{3\sqrt{22}}{2\sqrt{3}}$
26. $\sqrt{72} \times 3\sqrt{27}$
27. $4\sqrt{5} + 8\sqrt{45}$

Rewrite each of the following with rational (fractional) exponents.

28. $\sqrt[3]{2^4}$
29. $\sqrt[5]{1^2}$
30. $\sqrt[4]{8^3}$
31. $\sqrt[3]{6^2}$
32. $\sqrt[6]{2^{12}}$
33. $\sqrt[5]{9^3}$

Chapter 2 Test

1. Simplify: $\sqrt{135}$

   (A) $3\sqrt{15}$
   (B) $\sqrt{72}$
   (C) $9\sqrt{15}$
   (D) $\sqrt{9} \times \sqrt{15}$

2. Express $\dfrac{\sqrt{20}}{\sqrt{35}}$ in simplest form.

   (A) $\dfrac{2\sqrt{7}}{7}$
   (B) $\dfrac{2}{\sqrt{7}}$
   (C) $\dfrac{2\sqrt{5}}{\sqrt{7}}$
   (D) $\dfrac{4}{7}$

3. Simplify: $\dfrac{3\sqrt{12}}{2\sqrt{3}}$

   (A) 3
   (B) $\dfrac{3\sqrt{4}}{2}$
   (C) $\dfrac{6}{\sqrt{6}}$
   (D) $3\sqrt{3}$

4. Simplify: $\sqrt{44} \cdot 2\sqrt{33}$

   (A) $2\sqrt{77}$
   (B) $44\sqrt{3}$
   (C) $22\sqrt{7}$
   (D) $22\sqrt{12}$

5. Rewrite with rational exponents: $\sqrt[5]{3^2}$

   (A) 3^{10}
   (B) $5^{\frac{2}{3}}$
   (C) $3^{\frac{5}{2}}$
   (D) $3^{\frac{2}{5}}$

6. Simplify: $\sqrt{45} \times \sqrt{27}$

   (A) $3\sqrt{15}$
   (B) $\sqrt{72}$
   (C) $9\sqrt{15}$
   (D) $\sqrt{9} \times \sqrt{15}$

7. Simplify the following by rationalizing the denominator.

   $\dfrac{\sqrt{3}}{\sqrt{15}}$

   (A) $\dfrac{\sqrt{5}}{5}$
   (B) $\dfrac{1}{\sqrt{5}}$
   (C) $\dfrac{\sqrt{45}}{15}$
   (D) $\dfrac{3\sqrt{5}}{15}$

8. Simplify: $\sqrt[3]{40}$

   (A) $8\sqrt[3]{5}$
   (B) $\sqrt[3]{8} \times \sqrt[3]{5}$
   (C) $2\sqrt[3]{5}$
   (D) You cannot take the cube root of 40.

Chapter 3
Introduction to Algebra

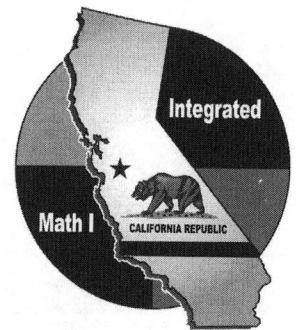

This chapter covers the following CA Integrated Math I standards:

| Algebra | 1.1 |

3.1 Algebra Vocabulary

Vocabulary Word	Example	Definition
variable	$4x$ (x is the variable)	a letter that can be replaced by a number
coefficient	$4x$ (4 is the coefficient)	a number multiplied by a variable or variables
term	$5x^2 + x - 2$ ($5x^2$, x, and -2 are terms)	numbers or variables separated by $+$ or $-$ signs
constant	$5x + 2y + 4$ (4 is a constant)	a term that does not have a variable
degree	$4x^2 + 3x - 2$ (the degree is 2)	the largest power of a variable in an expression
leading coefficient	$4x^2 + 3x - 2$ (4 is the leading coefficient)	the number multiplied by the term with the highest power
sentence	$2x = 7$ or $5 \leq x$	two algebraic expressions connected by $=, \neq, <, >, \leq, \geq$, or $\approx$
equation	$4x = 8$	a sentence with an equal sign
inequality	$7x < 30$ or $x \neq 6$	a sentence with one of the following signs: $\neq, <, >, \leq, \geq$, or $\approx$
base	6^3 (6 is the base)	the number used as a factor
exponent	6^3 (3 is the exponent)	the number of times the base is multiplied by itself

3.2 Substituting Numbers for Variables

These problems may look difficult at first glance, but they are very easy. Simply replace the variable with the number the variable is equal to, and solve the problems.

Example 1: In the following problems, substitute 10 for a.

Problem	Calculation	Solution
1. $a + 1$	Simply replace the a with 10. $10 + 1$	11
2. $17 - a$	$17 - 10$	7
3. $9a$	This means multiply. 9×10	90
4. $\dfrac{30}{a}$	This means divide. $30 \div 10$	3
5. a^3	$10 \times 10 \times 10$	1000
6. $5a + 6$	$(5 \times 10) + 6$	56

Note: Be sure to do all multiplying and dividing before adding and subtracting.

Example 2: In the following problems, let $x = 2$, $y = 4$, and $z = 5$.

Problem	Calculation	Solution
1. $5xy + z$	$5 \times 2 \times 4 + 5$	45
2. $xz^2 + 5$	$2 \times 5^2 + 5 = 2 \times 25 + 5$	55
3. $\dfrac{yz}{x}$	$(4 \times 5) \div 2 = 20 \div 2$	10

In the following problems, $t = 7$. Solve the problems.

1. $t + 3 =$
2. $18 - t =$
3. $\dfrac{21}{t} =$
4. $3t - 5 =$
5. $t^2 + 1 =$
6. $2t - 4 =$
7. $9t \div 3 =$
8. $\dfrac{t^2}{7} =$
9. $5t + 6 =$
10. $\dfrac{(t^2 - 7)}{6} =$
11. $4t + 5t =$
12. $\dfrac{6t}{3} =$

In the following problems $a = 4$, $b = -2$, $c = 5$, and $d = 10$. Solve the problems.

13. $4a + 2c =$
14. $3bc - d =$
15. $\dfrac{ac}{d} =$
16. $d - 2a =$
17. $a^2 - b =$
18. $abd =$
19. $5c - ad =$
20. $cd + bc =$
21. $\dfrac{6b}{a} =$
22. $9a + b =$
23. $5 + 3bc =$
24. $d^2 + d + 1 =$

Chapter 3   Introduction to Algebra

3.3   Understanding Algebra Word Problems

The biggest challenge to solving word problems is figuring out whether to add, subtract, multiply, or divide. Below is a list of key words and their meanings. This list does not include every situation you might see, but it includes the most common examples.

Words Indicating Addition	**Example**	**Add**
and	6 **and** 8	$6 + 8$
increased	The original price of $15 **increased** by $5.	$15 + 5$
more	3 coins and 8 **more**	$3 + 8$
more than	Josh has 10 points. Will has 5 **more than** Josh.	$10 + 5$
plus	8 baseballs **plus** 4 baseballs	$8 + 4$
sum	the **sum** of 3 and 5	$3 + 5$
total	the **total** of 10, 14, and 15	$10 + 14 + 15$

Words Indicating Subtraction	**Example**	**Subtract**
decreased	$16 **decreased** by $5	$16 - 5$
difference	the **difference** between 18 and 6	$18 - 6$
less	14 days **less** 5	$14 - 5$
less than	Jose completed 2 laps **less than** Mike's 9.	*$9 - 2$
left	Ray sold 15 out of 35 tickets. How many did he have **left**?	*$35 - 15$
lower than	This month's rainfall is 2 inches **lower than** last month's rainfall of 8 inches.	*$8 - 2$
minus	15 **minus** 6	$15 - 6$

* In subtraction word problems, you cannot always subtract the numbers in the order that they appear in the problem. Sometimes the first number should be subtracted from the last. You must read each problem carefully.

Words Indicating Multiplication	**Example**	**Multiply**
double	Her $1,000 profit **doubled** in in a month.	$1,000 \times 2$
half	**Half** of the $600 collected went to charity.	$\frac{1}{2} \times 600$
product	the **product** of 4 and 8	4×8
times	Li scored 3 **times** as many points as Ted who only scored 4.	3×4
triple	The bacteria **tripled** its original colony of 10,000 in just one day.	$3 \times 10,000$
twice	Ron has 6 CDs. Tom has **twice** as many.	2×6

Words Indicating Division	**Example**	**Divide**
divide into, by, or among	The group of 70 **divided into** 10 teams	$70 \div 10$ or $\frac{70}{10}$
quotient	the **quotient** of 30 and 6	$30 \div 6$ or $\frac{30}{6}$

3.3 Understanding Algebra Word Problems

Match the phrase with the correct algebraic expression below. The answers will be used more than once.

A. $y - 2$

B. $2y$

C. $y + 2$

D. $\dfrac{y}{2}$

E. $2 - y$

1. 2 more than y
2. y divided into 2
3. 2 less than y
4. twice y
5. the quotient of y and 2
6. y increased by 2
7. 2 less y
8. the product of 2 and y
9. y decreased by 2
10. y doubled
11. 2 minus y
12. the total of 2 and y

Now practice writing parts of algebraic expressions from the following word problems.

Example 3: the product of 3 and a number, t Answer: $3t$

13. 3 less than x
14. y divided among 10
15. the sum of t and 5
16. n minus 14
17. 5 times k
18. the total of z and 12
19. double the number b
20. x increased by 1
21. the quotient of t and 4
22. half of a number, y
23. bacteria culture, b, doubled
24. triple John's age, y
25. a number, n, plus 4
26. quantity, t, less 6
27. 18 divided by a number, x
28. n feet lower than 10
29. 3 more than p
30. the product of 4 and m
31. a number, y, decreased by 20
32. 5 times as much as x

3.4 Setting Up Algebra Word Problems

So far, you have seen only the first part of algebra word problems. To complete an algebra problem, an equal sign must be added. The words "**is**" or "**are**" as well as "**equal(s)**" signal that you should add an equal sign.

Example 4: Double Jake's age, x, minus 4 is 22.

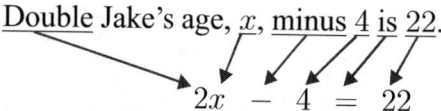

$$2x - 4 = 22$$

Translate the following word problems into algebra problems. DO NOT find the solutions to the problems yet.

1. Triple the original number, n, is $2{,}700$.

2. The product of a number, y, and 5 is equal to 15.

3. Four times the difference of a number, x, and 2 is 20.

4. The total, t, divided into 5 groups is 45.

5. The number of parts in inventory, p, minus 54 parts sold today is 320.

6. One-half an amount, x, added to $50 is $262

7. One hundred seeds divided by 5 rows equals n number of seeds per row.

8. A number, y, less than 50 is 82.

9. His base pay of $200 increased by his commission, x, is $500.

10. Seventeen more than half a number, h, is 35.

11. This month's sales of $2{,}300$ are double January's sales, x.

12. The quotient of a number, w, and 4 is 32.

13. Six less a number, d, is 12.

14. Four times the sum of a number, y, and 10 is 48.

15. We started with x number of students. When 5 moved away, we had 42 left.

16. A number, b, divided into 36 is 12.

3.5 Changing Algebra Word Problems to Algebraic Equations

Example 5: There are 3 people who have a total weight of 595 pounds. Sally weighs 20 pounds less than Jessie. Rafael weighs 15 pounds more than Jessie. How much does Jessie weigh?

Step 1: Notice everyone's weight is given in terms of Jessie. Sally weighs 20 pounds less than Jessie. Rafael weighs 15 pounds more than Jessie. First, we write everyone's weight in terms of Jessie, j.

$$\begin{aligned} \text{Jessie} &= j \\ \text{Sally} &= j - 20 \\ \text{Rafael} &= j + 15 \end{aligned}$$

Step 2: We know that all three together weigh 595 pounds. We write the sum of everyone's weight equal to 595.

$$j + j - 20 + j + 15 = 595$$

We will learn to solve these problems in Chapter 4.

Change the following word problems to algebraic equations.

1. Fluffy, Spot, and Shampy have a combined age in dog years of 91. Spot is 14 years younger than Fluffy. Shampy is 6 years older than Fluffy. What is Fluffy's age, f, in dog years?

2. Jerry Marcosi puts 5% of the amount he makes per week into a retirement account, r. He is paid $11.00 per hour and works 40 hours per week for a certain number of weeks, w. Write an equation to help him find out how much he puts into his retirement account.

3. A furniture store advertises a 40% off liquidation sale on all items. What would the sale price (p) be on a $2,530 dining room set?

4. Kyle Thornton buys an item which normally sells for a certain price, x. Today the item is selling for 25% off the regular price. A sales tax of 6% is added to the equation to find the final price, f.

5. Tamika Francois runs a floral shop. On Tuesday, Tamika sold a total of $600 worth of flowers. The flowers cost her $100, and she paid an employee to work 8 hours for a given wage, w. Write an equation to help Tamika find her profit, p, on Tuesday.

6. Sharice is a waitress at a local restaurant. She makes an hourly wage of $3.50, plus she receives tips. On Monday, she works 6 hours and receives tip money, t. Write an equation showing what Sharice makes on Monday, y.

7. Jenelle buys x shares of stock in a company at $34.50 per share. She later sells the shares at $40.50 per share. Write an equation to show how much money, m, Jenelle has made.

3.6 Substituting Numbers in Formulas

Example 6: Area of a parallelogram: $A = b \times h$
Find the area of the parallelogram if $b = 20$ cm and $h = 10$ cm.

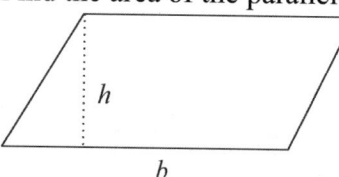

Step 1: Copy the formula with the numbers given in place of the letter in the formula.
$A = 20 \times 10$

Step 2: Solve the problem. $A = 20 \times 10 = 200$. Therefore, $A = 200$ cm².

Solve the following problems using the formulas given.

1. The volume of a rectangular pyramid is determined by using the following formula:
$V = \dfrac{lwh}{3}$
Find the volume of the pyramid if $l = 6$ in, $w = 6$ in, and $h = 11$ in.

2. Find the volume of a cone with a radius of 30 inches and a height of 60 inches using the formula:
$V = \frac{1}{3}\pi r^2 h \qquad \pi = 3.14$

3. Lumber is measured by the following formula:
Number of board feet $= \dfrac{LWT}{12}$
Find the number of board feet if $L = 14$ feet, $W = 8$ feet, and $T = 6$ feet.

4. The perimeter of a square is figured by the formula $P = 4s$.
Find the perimeter if $s = 6$.

5. What is the circumference of a circle with a diameter of 8 cm?
$C = \pi d \qquad \pi = 3.14$

6. Find the area of the trapezoid
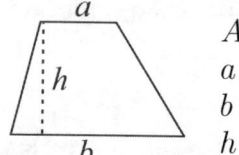
$A = \frac{1}{2}h(a+b)$
$a = 11$ in
$b = 23$ in
$h = 18$ in

7. Find the volume of a sphere with a radius of 6 cm. $\pi = 3.14$
$V = \frac{4}{3}\pi r^3$
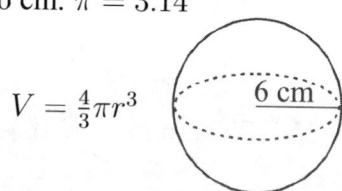

8. Find the area of the following ellipse given by the equation: $A = \pi ab$
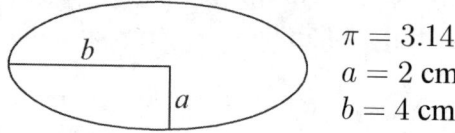
$\pi = 3.14$
$a = 2$ cm
$b = 4$ cm

9. The formula for changing from degrees Fahrenheit to degrees Celsius is:
$C = \dfrac{5(F - 32)}{9}$
If it is 68°F outside, how many degrees Celsius is it?

10. Find the volume. $V = \frac{4}{3}\pi r^3 \quad \pi = 3.14$
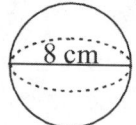

11. Louise has a cone-shaped mold to make candles. The diameter of the base is 10 cm, and it is 13 cm tall. How many cubic centimeters of liquid wax will it hold?
$\pi = 3.14$
$V = \frac{1}{3}\pi r^2 h$

3.7 Properties of Addition and Multiplication

The Associative, Commutative, and Distributive properties and the Identity of Addition and Multiplication are listed below by example as a quick refresher.

Property	Example
1. Associative Property of Addition	$(a + b) + c = a + (b + c)$
2. Associative Property of Multiplication	$(a \times b) \times c = a \times (b \times c)$
3. Commutative Property of Addition	$a + b = b + a$
4. Commutative Property of Multiplication	$a \times b = b \times a$
5. Distributive Property	$a \times (b + c) = (a \times b) + (a \times c)$
6. Identity Property of Addition	$0 + a = a$
7. Identity Property of Multiplication	$1 \times a = a$
8. Inverse Property of Addition	$a + (-a) = 0$
9. Inverse Property of Multiplication	$a \times \frac{1}{a} = \frac{a}{a} = 1, a \neq 0$

The Reflexive, Symmetric, and Transitive properties of equality are also listed with examples.

Property	Example
10. Reflexive Property of Equality	$a = a$
11. Symmetric Property of Equality	$a = b$ then $b = a$
12. Transitive Property of Equality (If the first number equals the second, and the second is equal to the third, then the first must be equal to the third.)	If $a = b$ and $b = c$ then $a = c$ $a \times (b + c) = (a \times b) + (a \times c)$

Write the number of the property listed above that describes each of the following statements.

1. $4 + 5 = 5 + 4$
2. $4 + (2 + 8) = (4 + 2) + 8$
3. $10(4 + 7) = (10)(4) + (10)(7)$
4. $(2 \times 3) \times 4 = 2 \times (3 \times 4)$
5. $1 \times 12 = 12$
6. $8 \left(\frac{1}{8} \right) = 1$
7. $1c = c$
8. If $t = z$ and $z = q$, then $t = q$
9. $42 = 42$
10. $18 + 0 = 18$
11. $9 + (-9) = 0$
12. $p \times q = q \times p$
13. $t + 0 = t$
14. $x(y + z) = xy + xz$
15. $(m)(n \cdot p) = (m \cdot n)(p)$
16. $-y + y = 0$
17. If $a = z$, then $z = a$
18. If $f = g$ and $g = 107$, then $f = 107$

Chapter 3 Review

Solve the following problems using $x = 2$.

1. $3x + 4 =$
2. $\dfrac{6x}{4} =$
3. $x^2 - 5 =$
4. $\dfrac{x^3 + 8}{2} =$
5. $12 - 3x =$
6. $x - 5 =$
7. $-5x + 4 =$
8. $9 - x =$
9. $2x + 2 =$

Solve the following problems. Let $w = -1, y = 3, z = 5$.

10. $5w - y =$
11. $wyz + 2 =$
12. $z - 2w =$
13. $\dfrac{3z + 5}{wz} =$
14. $\dfrac{6w}{y} + \dfrac{z}{w} =$
15. $25 - 2yz =$
16. $-2y + 3$
17. $4w - (yw) =$
18. $7y - 5z =$

Identify the property used in each equation below.

19. $a + 0 = a$
20. $1 \times a = a$
21. $x(y + z) = xy + xz$
22. $a\left(\dfrac{1}{a}\right) = 1$
23. $(m)(n \cdot p) = (m \cdot n)(p)$
24. $-y + y = 0$
25. If $m = n$ and $n = o$, then $m = o$.
26. $a \times b = b \times a$

For questions 27–29, write an equation to match the problem.

27. Calista earns $450 per week for a 40-hour work week plus $16.83 per hour for each hour of overtime after 40 hours. Write an equation that would be used to determine her weekly wages where w is her wages and v is the number of overtime hours worked.

28. Daniel purchased a 1-year CD, c, from a bank. He bought it at an annual interest rate of 6%. After 1 year, Daniel cashes in the CD. What is the total amount it is worth?

29. Omar is a salesman. He earns an hourly wage of $8.00 per hour, plus he receives a commission of 7% on the sales he makes. Write an equation which would be used to determine his weekly salary, w, where x is the number of hours worked, and y is the amount of sales for the week.

Chapter 3 Review

Answer each of the following questions.

30. Juan sells a boat that he bought 5 years ago. He sells it for 60% less than he originally paid for it. If the original cost is b, write an expression that shows how much he sells the boat for.

31. Toshi is going to get a 7% raise after he works at his job for 1 year. If s represents his starting salary, write an expression that shows how much he will make after his raise, x.

32. Lumber is measured with the following formula:

    Number of board feet $= \dfrac{LWT}{12}$

    $L = $ Length of the board in feet
    $W = $ Width of the board in feet
    $T = $ Thickness of the board in feet

    Find the number of board feet if $L = 12$ feet, $W = 6$ feet, and $T = \frac{1}{4}$ feet.

33. To convert from degrees Celsius to degrees Fahrenheit, use the following formula:

    $$F = \dfrac{9C}{5} + 32$$

    If it is 15°C outside, what is the temperature in degrees Fahrenheit?

Chapter 3   Introduction to Algebra

Chapter 3 Test

1. Solving the following expression using $a = 4$.

   $3a - 2$

   (A) 14
   (B) 10
   (C) 16
   (D) 12

2. Solving the following expression using $y = 3$.

   $8y \div 3$

   (A) 8
   (B) 21
   (C) 24
   (D) 72

3. Write the expression from the following word problem.

   A number divided by the sum of nine and two.

   (A) $\dfrac{x}{9+2}$
   (B) $\dfrac{9+2}{x}$
   (C) $\dfrac{x}{9-2}$
   (D) $\dfrac{x+2}{9}$

4. A box has length of 20 inches, the height of 12 inches, and the width of 38 inches. What is the volume of the box using the formula $V = lwh$?

   (A) 70 inches
   (B) 240 inches
   (C) 9,120 inches
   (D) 2,912 inches

5. Solving the following expression using $x = 2$ and $y = 5$.

   $3x + 4y - 1$

   (A) 22
   (B) 13
   (C) 25
   (D) 10

6. Write the expression from the following word problem.

   Five less than x plus seven.

   (A) $5 - (x + 7)$
   (B) $(x + 5) - 7$
   (C) $(x + 7) - 5$
   (D) $(x - 7) + 5$

7. Write the expression from the following word problem.

   Fifteen minus a number, then divided by two equals eleven.

   (A) $\dfrac{15 - y}{2} = 11$
   (B) $11 - \dfrac{y}{2} = 15$
   (C) $15 - \dfrac{y}{2} = 11$
   (D) $2 - \dfrac{y}{15} = 11$

8. Write the equation from the following word problem.

   Sixteen times two plus a number equals seven.

   (A) $16 \times 2 + x = 7$
   (B) $16 \times 2 = 7x$
   (C) $16 \times (2 + x) = 7$
   (D) $16 + 2 \times x = 7$

9. Which property is demonstrated by the expression below.
$2 + (3 + 8) = (2 + 3) + 8$

   (A) commutative property of addition
   (B) associative property of addition
   (C) distributive property
   (D) identity property of addition

10. Which property is demonstrated by the expression below.
If $x = y$ and $y = z$, then $x = z$.

    (A) reflexive property of equality
    (B) symmetric property of equality
    (C) transitive property of equality
    (D) distributive property

11. Is the expression 15×8 equivalent to the expression 8×15?

    (A) Yes because of the commutative property.
    (B) Yes because of the associative property.
    (C) Yes because of the distributive property.
    (D) Yes because of the inverse property.

12. Is the equation $5(x - 6) = -2$ equivalent to the equation $5x - 30 = -2$?

    (A) Yes because of the commutative property.
    (B) Yes because of the distributive property.
    (C) Yes because of the associative property.
    (D) Yes because of the inverse property.

13. Tom earns $500 per week before taxes are taken out. His employer takes out a total of 33% for state, federal, and Social Security taxes. Which expression below will help Tom figure his net pay?

    (A) $500 - 0.33$
    (B) $500 \div 0.33$
    (C) $500 + 0.33(500)$
    (D) $500 - 0.33(500)$

14. Rosa has to pay the first $100 of her medical expenses each year before she qualifies for her insurance company to begin paying. After paying the $100 "deductible," her insurance company will pay 80% of her medical expenses. This year, her total medical expenses came to $960.00. Which expression below shows how much her insurance company will pay?

    (A) $0.80(960 - 100)$
    (B) $100 + (960 \div 0.80)$
    (C) $960(100 - 0.80)$
    (D) $0.80(960 + 100)$

15. A plumber charges $45 per hour plus a $25.00 service charge. If a represents his total charges in dollars and b represents the number of hours worked, which formula below could the plumber use to calculate his total charges?

    (A) $a = 45 + 25b$
    (B) $a = 45 + 25 + b$
    (C) $a = 45b + 25$
    (D) $a = (45)(25) + b$

16. In 2007, Bell Computers informed its sales force to expect a 2.6% price increase on all computer equipment in the year 2008. A certain sales representative wanted to see how much the increase would be on a computer, c, that sold for $2200 in 2007. Which expression below will help him find the cost of the computer in the year 2008?

    (A) $0.26(2200)$
    (B) $2200 - 0.026(2200)$
    (C) $2200 + 0.026(2200)$
    (D) $0.026(2200) - 2200$

Chapter 4
Solving Multi-Step Equations and Inequalities

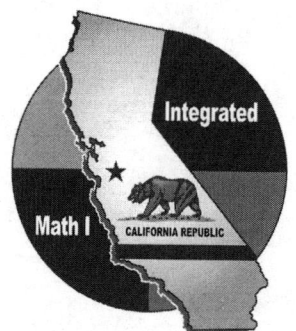

This chapter covers the following CA Integrated Math I standards:

Algebra	2.0
	4.0
	25.2

4.1 Two-Step Algebra Problems

In the following two-step algebra problems, **additions** and **subtractions** are performed first and then **multiplication** and **division**.

Example 1: $-4x + 7 = 31$

Step 1: Subtract 7 from both sides.

$$\begin{array}{r} -4x + 7 = 31 \\ -7 \quad -7 \\ \hline -4x = 24 \end{array}$$

Step 2: Divide both sides by -4.

$$\frac{-4x}{-4} = \frac{24}{-4} \quad \text{so } x = -6$$

Example 2: $-8 - y = 12$

Step 1: Add 8 to both sides.

$$\begin{array}{r} -8 - y = 12 \\ +8 \quad +8 \\ \hline -y = 20 \end{array}$$

Step 2: To finish solving a problem with a negative sign in front of the variable, multiply both sides by -1. The variable needs to be positive in the answer.

$$(-1)(-y) = (-1)(20) \text{ so } y = -20$$

Solve the two-step algebra problems below.

1. $6x - 4 = -34$
2. $5y - 3 = 32$
3. $8 - t = 1$
4. $10p - 6 = -36$
5. $11 - 9m = -70$
6. $4x - 12 = 24$
7. $3x - 17 = -41$
8. $9d - 5 = 49$
9. $10h + 8 = 78$
10. $-6b - 8 = 10$
11. $-g - 24 = -17$
12. $-7k - 12 = 30$
13. $9 - 5r = 64$
14. $6y - 14 = 34$
15. $12f + 15 = 51$
16. $21t + 17 = 80$
17. $20y + 9 = 149$
18. $15p - 27 = 33$
19. $22h + 9 = 97$
20. $-5 + 36w = 175$

4.2 Two-Step Algebra Problems with Fractions

An algebra problem may contain a fraction. Study the following example to understand how to solve algebra problems that contain a fraction.

Example 3: $\frac{x}{2} + 4 = 3$

Step 1:
$$\frac{x}{2} + 4 = 3$$
$$\underline{-4 -4}$$
$$\frac{x}{2} = -1$$
Subtract 4 from both sides.

Step 2: $\frac{x}{2} = -1$ Multiply both sides by 2 to eliminate the fraction.

$$\frac{x}{\cancel{2}} \times \cancel{2} = -1 \times 2, \; x = -2$$

Simplify the following algebra problems.

1. $4 + \frac{y}{3} = 7$
2. $\frac{a}{2} + 5 = 12$
3. $\frac{w}{5} - 3 = 6$
4. $\frac{x}{9} - 9 = -5$
5. $\frac{b}{6} + 2 = -4$
6. $7 + \frac{z}{2} = -13$
7. $\frac{x}{2} - 7 = 3$
8. $\frac{c}{5} + 6 = -2$
9. $3 + \frac{x}{11} = 7$
10. $16 + \frac{m}{6} = 14$
11. $\frac{p}{3} + 5 = -2$
12. $\frac{t}{8} + 9 = 3$
13. $\frac{v}{7} - 8 = -1$
14. $5 + \frac{h}{10} = 8$
15. $\frac{k}{7} - 9 = 1$
16. $\frac{y}{4} + 13 = 8$
17. $15 + \frac{z}{14} = 13$
18. $\frac{b}{6} - 9 = -14$
19. $\frac{d}{3} + 7 = 12$
20. $10 + \frac{b}{6} = 4$
21. $2 + \frac{p}{4} = -6$
22. $\frac{t}{7} - 9 = -5$
23. $\frac{a}{10} - 1 = 3$
24. $\frac{a}{8} + 16 = 9$

Chapter 4 Solving Multi-Step Equations and Inequalities

4.3 More Two-Step Algebra Problems with Fractions

Study the following example to understand how to solve algebra problems that contain a different type of fraction.

Example 4: $\dfrac{x+2}{4} = 3$ In this example, "$x + 2$" is divided by 4, and not just the x or the 2.

Step 1: $\dfrac{x+2}{4} \times 4 = 3 \times 4$ First multiply both sides by 4 to eliminate the fraction.

Step 2:
$$\begin{array}{rl} x + 2 &= 12 \\ -2 & -2 \\ \hline x &= 10 \end{array}$$
Next, subtract 2 from both sides.

Solve the following problems.

1. $\dfrac{x+1}{5} = 4$

2. $\dfrac{z-9}{2} = 7$

3. $\dfrac{b-4}{4} = -5$

4. $\dfrac{y-9}{3} = 7$

5. $\dfrac{d-10}{-2} = 12$

6. $\dfrac{w-10}{-8} = -4$

7. $\dfrac{x-1}{-2} = -5$

8. $\dfrac{c+40}{-5} = -7$

9. $\dfrac{13+h}{2} = 12$

10. $\dfrac{k-10}{3} = 9$

11. $\dfrac{a+11}{-4} = 4$

12. $\dfrac{x-20}{7} = 6$

13. $\dfrac{t+2}{6} = -5$

14. $\dfrac{b+1}{-7} = 2$

15. $\dfrac{f-9}{3} = 8$

16. $\dfrac{4+w}{6} = -6$

17. $\dfrac{3+t}{3} = 10$

18. $\dfrac{x+5}{5} = -3$

19. $\dfrac{g+3}{2} = 11$

20. $\dfrac{k+1}{-6} = 5$

21. $\dfrac{y-14}{2} = -8$

22. $\dfrac{z-4}{-2} = 13$

23. $\dfrac{w+2}{15} = -1$

24. $\dfrac{3+h}{3} = 6$

4.4 Combining Like Terms

In algebra problems, separate **terms** by $+$ and $-$ signs. The expression $5x - 4 - 3x + 7$ has 4 terms: $5x$, 4, $3x$, and 7. Terms having the same variable can be combined (added or subtracted) to simplify the expression. $5x - 4 - 3x + 7$ simplifies to $2x + 3$.

$$5x - 3x \quad -4 + 7 = 2x + 3$$

Simplify the following expressions.

1. $7x + 12x$
2. $8y - 5y + 8$
3. $4 - 2x + 9$
4. $11a - 16 - a$
5. $9w + 3w + 3$
6. $-5x + x + 2x$
7. $w - 15 + 9w$
8. $21 - 10t + 9 - 2t$
9. $-3 + x - 4x + 9$
10. $7b + 12 + 4b$
11. $4h - h + 2 - 5$
12. $-6k + 10 - 4k$
13. $2a + 12a - 5 + a$
14. $5 + 9c - 10$
15. $-d + 1 + 2d - 4$
16. $-8 + 4h + 1 - h$
17. $12x - 4x + 7$
18. $10 + 3z + z - 5$
19. $14 + 3y - y - 2$
20. $11p - 4 + p$
21. $11m + 2 - m + 1$

4.5 Solving Equations with Like Terms

When an equation has two or more like terms on the same side of the equation, combine like terms as the **first** step in solving the equation.

Example 5: $\quad 7x + 2x - 7 = 21 + 8$

Step 1: Combine like terms on both sides of the equation.

Step 2: Solve the two-step algebra problem as explained previously.

$$\begin{aligned} 7x + 2x - 7 &= 21 + 8 \\ 9x - 7 &= 29 \\ +7 &\quad +7 \\ 9x \div 9 &= 36 \div 9 \\ x &= 4 \end{aligned}$$

Solve the equations below combining like terms first.

1. $3w - 2w + 4 = 6$
2. $7x + 3 + x = 16 + 3$
3. $5 - 6y + 9y = -15 + 5$
4. $-14 + 7a + 2a = -5$
5. $-2t + 4t - 7 = 9$
6. $9d + d - 3d = 14$
7. $-6c - 4 - 5c = 10 + 8$
8. $15m - 9 - 6m = 9$
9. $-4 - 3x - x = -16$
10. $9 - 12p + 5p = 14 + 2$
11. $10y + 4 - 7y = -17$
12. $-8a - 15 - 4a = 9$

Chapter 4 Solving Multi-Step Equations and Inequalities

If the equation has like terms on both sides of the equation, you must get all of the terms with a **variable** on one side of the equation and all of the **integers** on the other side of the equation.

Example 6: $3x + 2 = 6x - 1$

Step 1: Subtract $6x$ from both sides to move all the **variables** to the left side.

Step 2: Subtract 2 from both sides to move all the **integers** to the right side.

Step 3: Divide by -3 to solve for x.

$$\begin{aligned} 3x + 2 &= 6x - 1 \\ -6x & \quad -6x \\ \hline -3x + 2 &= -1 \\ -2 & \quad -2 \\ \hline \frac{-3x}{-3} &= \frac{-3}{-3} \\ x &= 1 \end{aligned}$$

Solve the following problems.

1. $3a + 1 = a + 9$
2. $2d - 12 = d + 3$
3. $5x + 6 = 14 - 3x$
4. $15 - 4y = 2y - 3$
5. $9w - 7 = 12w - 13$
6. $10b + 19 = 4b - 5$
7. $-7m + 9 = 29 - 2m$
8. $5x - 26 = 13x - 2$
9. $19 - p = 3p - 9$
10. $-7p - 14 = -2p + 11$

11. $16y + 12 = 9y + 33$
12. $13 - 11w = 3 - w$
13. $-17b + 23 = -4 - 8b$
14. $k + 5 = 20 - 2k$
15. $12 + m = 4m + 21$
16. $7p - 30 = p + 6$
17. $19 - 13z = 9 - 12z$
18. $8y - 2 = 4y + 22$
19. $5 + 16w = 6w - 45$
20. $-27 - 7x = 2x + 18$

21. $-12x + 14 = 8x - 46$
22. $27 - 11h = 5 - 9h$
23. $5t + 36 = -6 - 2t$
24. $17y + 42 = 10y + 7$
25. $22x - 24 = 14x - 8$
26. $p - 1 = 4p + 17$
27. $4d + 14 = 3d - 1$
28. $7w - 5 = 8w + 12$
29. $-3y - 2 = 9y + 22$
30. $17 - 9m = m - 23$

4.6 Removing Parentheses

The distributive principle is used to remove parentheses.

Example 7: $2(a+6)$
You multiply 2 by each term inside the parentheses. $2 \times a = 2a$ and $2 \times 6 = 12$. The 12 is a positive number so use a plus sign between the terms in the answer.
$2(a+6) = 2a + 12$

Example 8: $4(-5c + 2)$
The first term inside the parentheses could be negative. Multiply in exactly the same way as the examples above. $4 \times (-5c) = -20c$ and $4 \times 2 = 8$
$4(-5c + 2) = -20c + 8$

Remove the parentheses in the problems below.

1. $7(n+6)$
2. $8(2g-5)$
3. $11(5z-2)$
4. $6(-y-4)$
5. $3(-3k+5)$
6. $4(d-8)$
7. $2(-4x+6)$
8. $7(4+6p)$
9. $5(-4w-8)$
10. $6(11x+2)$
11. $10(9-y)$
12. $9(c-9)$
13. $12(-3t+1)$
14. $3(4y+9)$
15. $8(b+3)$

The number in front of the parentheses can also be negative. Remove these parentheses the same way.

Example 9: $-2(b-4)$
First, multiply $-2 \times b = -2b$
Second, multiply $-2 \times -4 = 8$
Copy the two products. The second product is a positive number so put a plus sign between the terms in the answer.
$-2(b-4) = -2b + 8$

Remove the parentheses in the following problems.

16. $-7(x+2)$
17. $-5(4-y)$
18. $-4(2b-2)$
19. $-2(8c+6)$
20. $-5(-w-8)$
21. $-3(4x-2)$
22. $-2(-z+2)$
23. $-4(7p+7)$
24. $-9(t-6)$
25. $-10(2w+4)$
26. $-3(9-7p)$
27. $-9(-k-3)$
28. $-1(7b-9)$
29. $-6(-5t-2)$
30. $-7(-v+4)$

Chapter 4   Solving Multi-Step Equations and Inequalities

4.7   Multi-Step Algebra Problems

You can now use what you know about removing parentheses, combining like terms, and solving simple algebra problems to solve problems that involve three or more steps. Study the examples below to see how easy it is to solve multi-step problems.

Example 10: $3(x+6) = 5x - 2$

   Step 1:   Use the distributive property to remove parentheses.

   Step 2:   Subtract $5x$ from each side to move the terms with variables to the left side of the equation.

   Step 3:   Subtract 18 from each side to move the integers to the right side of the equation.

   Step 4:   Divide both sides by -2 to solve for x.

$$\begin{aligned} 3x + 18 &= 5x - 2 \\ -5x & \quad -5x \\ \hline -2x + 18 &= -2 \\ -18 & \quad -18 \\ \hline \frac{-2x}{-2} &= \frac{-20}{-2} \\ x &= 10 \end{aligned}$$

Example 11: $\dfrac{3(x-3)}{2} = 9$

   Step 1:   Use the distributive property to remove parentheses.

   Step 2:   Multiply both sides by 2 to eliminate the fraction.

   Step 3:   Add 9 to both sides, and combine like terms.

   Step 4:   Divide both sides by 3 to solve for x.

$$\begin{aligned} \frac{3x - 9}{2} &= 9 \\ \frac{2(3x-9)}{2} &= 2(9) \\ 3x - 9 &= 18 \\ +9 & \quad +9 \\ \hline \frac{3x}{3} &= \frac{27}{3} \\ x &= 9 \end{aligned}$$

Solve the following multi-step algebra problems.

1. $2(y-3) = 4y + 6$

2. $\dfrac{2(a+4)}{2} = 12$

3. $\dfrac{10(x-2)}{5} = 14$

4. $\dfrac{12y - 18}{6} = 4y + 3$

5. $2x + 3x = 30 - x$

6. $\dfrac{2a+1}{3} = a + 5$

7. $5(b-4) = 10b + 5$

8. $-8(y+4) = 10y + 4$

4.7 Multi-Step Algebra Problems

9. $\dfrac{x+4}{-3} = 6 - x$

10. $\dfrac{4(n+3)}{5} = n - 3$

11. $3(2x - 5) = 8x - 9$

12. $7 - 10a = 9 - 9a$

13. $7 - 5x = 10 - (6x + 7)$

14. $4(x - 3) - x = x - 6$

15. $4a + 4 = 3a - 4$

16. $-3(x - 4) + 5 = -2x - 2$

17. $5b - 11 = 13 - b$

18. $\dfrac{-4x + 3}{2x} = \dfrac{7}{2x}$

19. $-(x + 1) = -2(5 - x)$

20. $4(2c + 3) - 7 = 13$

21. $6 - 3a = 9 - 2(2a + 5)$

22. $-5x + 9 = -3x + 11$

23. $3y + 2 - 2y - 5 = 4y + 3$

24. $3y - 10 = 4 - 4y$

25. $-(a + 3) = -2(2a + 1) - 7$

26. $5m - 2(m + 1) = m - 10$

27. $\dfrac{1}{2}(b - 2) = 5$

28. $-3(b - 4) = -2b$

29. $4x + 12 = -2(x + 3)$

30. $\dfrac{7x + 4}{3} = 2x - 1$

31. $9x - 5 = 8x - 7$

32. $7x - 5 = 4x + 10$

33. $\dfrac{4x + 8}{2} = 6$

34. $2(c + 4) + 8 = 10$

35. $y - (y + 3) = y + 6$

36. $4 + x - 2(x - 6) = 8$

Chapter 4 Solving Multi-Step Equations and Inequalities

4.8 Multi-Step Inequalities

Remember that adding and subtracting with inequalities follow the same rules as equations. When you multiply or divide both sides of an inequality by the same positive number, the rules are also the same as for equations. However, when you multiply or divide both sides of an inequality by a **negative** number, you must **reverse** the inequality symbol.

Example 12:
$$-x > 4$$
$$(-1)(-x) < (-1)(4)$$
$$x < -4$$

Example 13:
$$-4x < 2$$
$$\frac{-4x}{-4} > \frac{2}{-4}$$
$$x > -\frac{1}{2}$$

Reverse the symbol when you multiply or divide by a negative number.

When solving multi-step inequalities, first add and subtract to isolate the term with the variable. Then multiply and divide.

Example 14: $2x - 8 > 4x + 1$

Step 1: Add 8 to both sides.

$$2x - 8 + 8 > 4x + 1 + 8$$
$$2x > 4x + 9$$

Step 2: Subtract $4x$ from both sides.

$$2x - 4x > 4x + 9 - 4x$$
$$-2x > 9$$

Step 3: Divide by -2. Remember to change the direction of the inequality sign.

$$\frac{-2x}{-2} < \frac{9}{-2}$$
$$x < -\frac{9}{2}$$

54 Copyright © American Book Company

4.8 Multi-Step Inequalities

Solve each of the following inequalities.

1. $8 - 3x \leq 7x - 2$

2. $3(2x - 5) \geq 8x - 5$

3. $\frac{1}{3}b - 2 > 5$

4. $7 + 3y > 2y - 5$

5. $3a + 5 < 2a - 6$

6. $3(a - 2) > -5a - 2(3 - a)$

7. $2x - 7 \geq 4(x - 3) + 3x$

8. $6x - 2 \leq 5x + 5$

9. $-\frac{x}{4} > 12$

10. $-\frac{2x}{3} \leq 6$

11. $3b + 5 < 2b - 8$

12. $4x - 5 \leq 7x + 13$

13. $4x + 5 \leq -2$

14. $2y - 5 > 7$

15. $4 + 2(3 - 2y) \leq 6y - 20$

16. $-4c + 6 \leq 8$

17. $-\frac{1}{2}x + 2 > 9$

18. $\frac{1}{4}y - 3 \leq 1$

19. $-3x + 4 > 5$

20. $\frac{y}{2} - 2 \geq 10$

21. $7 + 4c < -2$

22. $2 - \frac{a}{2} > 1$

23. $10 + 4b \leq -2$

24. $-\frac{1}{2}x + 3 > 4$

Chapter 4   Solving Multi-Step Equations and Inequalities

Chapter 4 Review

Solve each of the following equations.

1. $4a - 8 = 28$

2. $5 + \dfrac{x}{8} = -4$

3. $-7 + 23w = 108$

4. $\dfrac{y-8}{6} = 7$

5. $c - 13 = 5$

6. $\dfrac{b+9}{12} = -3$

7. $19 - 8d = d - 17$

8. $\dfrac{-3}{x} + 11 = -1$

9. $7w - 8w = -4w - 30$

10. $\dfrac{12}{f} - 7 = -5$

11. $6 + 16x = -2x - 12$

12. $6 - \dfrac{1}{q} = 4$

Remove parentheses.

13. $3(-4x + 7)$

14. $11(2y + 5)$

15. $6(8 - 9b)$

16. $-8(-2 + 3a)$

17. $-2(5c - 3)$

18. $-5(7y - 1)$

Solve each of the following equations and inequalities.

19. $\dfrac{-11c - 35}{4} = 4c - 2$

20. $5 + x - 3(x + 4) = -17$

21. $4(2x + 3) \geq 2x$

22. $7 - 3x \leq 6x - 2$

23. $\dfrac{5(n+4)}{3} = n - 8$

24. $-y > 14$

25. $2(3x - 1) \geq 3x - 7$

26. $3(x + 2) < 7x - 10$

Chapter 4 Test

1. Find the value of n. $19n - 57 = 76$

   (A) 1
   (B) 3
   (C) 5
   (D) 7

2. Solve for x. $14x + 84 = 154$

   (A) 4
   (B) 5
   (C) 11
   (D) 17

3. Which of the following is equivalent to $4 - 5x > 3(x - 4)$?

   (A) $4 - 5x > 3x - 4$
   (B) $4 - 5x > 3x - 12$
   (C) $4 - 5x > 3x - 1$
   (D) $4 - 5x > 3x - 7$

4. Which of the following is equivalent to $3(x - 2) + 1 - 2x = -4$?

   (A) $x - 6 = -4$
   (B) $-6x + 1 = -4$
   (C) $5x - 7 = -4$
   (D) $x - 5 = -4$

5. $5(2x + 11) - 3 \times 5 =$

   (A) $7x + 40$
   (B) $7x + 20$
   (C) $10x + 40$
   (D) $10x + 260$

6. Solve for b. $4b - 8 < 56$

   (A) $b < 12$
   (B) $b < 16$
   (C) $b < -12$
   (D) $b < -16$

7. Which of the following is equivalent to $3(2x - 5) - 4(x - 3) = 7$?

   (A) $x + 27 = 7$
   (B) $2x - 3 = 7$
   (C) $10x - 27 = 7$
   (D) $x - 27 = 7$

8. $3(x - 2) - 1 = 6(x + 5)$

   (A) -4
   (B) $-\dfrac{37}{3}$
   (C) 4
   (D) $\dfrac{23}{3}$

Chapter 5
Algebra Word Problems

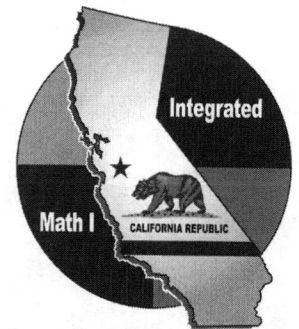

This chapter covers the following CA Integrated Math I standards:

| Algebra | 5.0 |

An equation states that two mathematical expressions are equal. In working with word problems, the words that mean equal are **equals, is, was, is equal to, amounts to,** and other expressions with the same meaning. To translate a word problem into an algebraic equation, use a variable to represent the unknown or unknowns you are looking for. In the following example, let n be the number you are looking for.

Example 1: Four more than twice a number is two less than three times the number.

Step 1: **Translation:**
$$4 + 2n = 3n - 2$$

Step 2: **Now Solve:**
$$\begin{array}{rcl} -2n & & -2n \\ \hline 4 & = & n - 2 \\ +2 & & +2 \\ \hline 6 & = & n \end{array}$$

The number is 6.
Substitute the number back into the original equation to check.

Translate the following word problems into equations and solve.

1. Four less than twice a number is ten. Find the number.

2. Three more than three times a number is one less than two times the number. What is the number?

3. The sum of seven times a number and the number is 24. What is the number?

4. Negative 18 is the sum of five and a number. Find the number.

5. Negative 14 is equal to ten minus the product of six and a number. What is the number?

6. Two less than twice a number equals the number plus 12. What is the number?

7. The difference between three times a number and 31 is two. What is the number?

8. Sixteen is fourteen less than the product of a number and five. What is the number?

9. Eight more than twice a number is four times the difference between five and the number. What is the number?

10. Three less than twice a number is three times the sum of one and the number. What is the number?

5.1 Geometry Word Problems

The perimeter of a geometric figure is the distance around the outside of the figure.

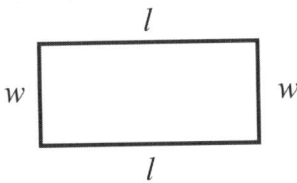

perimeter = $2l + 2w$

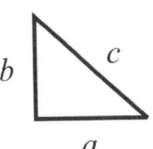

perimeter = $a + b + c$

Example 2: The perimeter of a rectangle is 44 feet. The length of the rectangle is 6 feet more than the width. What is the measure of the width?

Step 1: Let the variable be the length of the unknown side.
width = w length = $6 + w$

Step 2: Use the equation for the perimeter of a rectangle as follows:
$2l + 2w$ = perimeter of a rectangle
$2(w + 6) + 2w = 44$

Step 3: Solve for w.

Solution: width = 8 feet

Example 3: The perimeter of a triangle is 26 feet. The second side is twice as long as the first. The third side is 1 foot longer than the second side. What is the length of the 3 sides?

Step 1: Let x = first side $2x$ = second side $2x + 1$ = third side

Step 2: Use the equation for perimeter of a triangle as follows:
sum of the length of the sides = perimeter of a triangle.
$x + 2x + 2x + 1 = 26$

Step 3: Solve for x. $5x + 1 = 26$ so $x = 5$

Solution: first side $x = 5$ second side $2x = 10$ third side $2x + 1 = 11$

Solve the following word problems.

1. The length of a rectangle is 4 times longer than the width. The perimeter is 30. What is the width?

2. The length of a rectangle is 3 more than twice the width. The perimeter is 36. What is the length?

3. The perimeter of a triangle is 18 feet. The second side is two feet longer than the first. The third side is two feet longer then the second. What are the lengths of the sides?

4. In an isosceles triangle, two sides are equal. The third side is two less than twice the length of the sum of the two sides. The perimeter is 40. What are the lengths of the three sides?

5. The sum of the measures of the angles of a triangle is 180°. The second angle is three times the measure of the first angle. The third angle is four times the measure of the second angle. Find the measure of each angle.

6. The sum of the measures of the angles of a triangle is 180°. The second angle of a triangle is twice the measure of the first angle. The third angle is 20 more than 5 times the first. What are the measures of the three angles?

5.2 Age Problems

Example 4: Tara is twice as old as Gwen. Their sister, Amy, is 5 years older than Gwen. If the sum of their ages is 29 years, find each of their ages.

Step 1: We want to find each of their ages so there are three unknowns. Tara is twice as old as Gwen, and Amy is older than Gwen, so Gwen is the youngest. Let x be Gwen's age. From the problem we can see that:

$$\left. \begin{array}{rcl} \text{Gwen} &=& x \\ \text{Tara} &=& 2x \\ \text{Amy} &=& x+5 \end{array} \right\} \text{The sum of their ages is 29.}$$

Step 2: Set up the equation, and solve for x.

$$\begin{aligned} x + 2x + x + 5 &= 29 \\ 4x + 5 &= 29 \\ 4x &= 29 - 5 \\ x &= \frac{24}{4} \\ x &= 6 \end{aligned}$$

Solution: Gwen's age $(x) = 6$
Tara's age $(2x) = 12$
Amy's age $(x+5) = 11$

Solve the following age problems.

1. Carol is 15 years older than her cousin Amanda. Cousin Bill is 4 times as old as Amanda. The sum of their ages is 99. Find each of their ages.

2. Derrick is 5 less than twice as old as Brandon. The sum of their ages is 31. How old are Derrick and Brandon?

3. Beth's mom is 5 times older than Beth. Beth's dad is 8 years older than Beth's mom. The sum of their ages is 74. How old are each of them?

5.2 Age Problems

4. Delores is 4 years more than three times as old as her son, Raul. If the difference between their ages is 34, how old are Delores and Raul?

5. Eileen is 9 years older than Karen. John is three times as old as Karen. The sum of their ages is 64. How old are Eileen, Karen, and John?

6. Taylor is 20 years younger than Jim. Andrew is twice as old as Taylor. The sum of their ages is 32. How old are Taylor, Jim, and Andrew?

The following problems work in the same way as the age problems. There are two or three items of different weight, distance, number, or size. You are given the total and asked to find the amount of each item.

7. Three boxes have a total height of 720 pounds. Box A weighs twice as much as Box B. Box C weighs 30 pounds more than Box A. How much do each of the boxes weigh?

8. There are 170 students registered for American History classes. There are twice as many students registered in second period as first period. There are 10 less than three times as many students registered in third period as in first period. How many students are in each period?

9. Mei earns $4 less than three times as much as Olivia. Shane earns twice as much as Mei. Together they earn $468 per week. How much does each person earn per week?

10. Ellie, the elephant, eats 4 times as much as Popcorn, the pony. Zac, the zebra, eats twice as much as Popcorn. Altogether, they eat 238 kilograms of feed per week. How much feed does each of them require each week?

11. The school cafeteria served three kinds of lunches today to 117 students. The students chose the cheeseburgers three times more often than the grilled cheese sandwiches. There were twice as many grilled cheese sandwiches sold as fish sandwiches. How many of each lunch were served?

12. Three friends drove southeast to New Mexico. Kyle drove half as far as Jamaal. Conner drove 4 times as far as Kyle. Altogether, they drove 476 miles. How far did each friend drive?

13. Bianca is taking collections for this year's Feed the Hungry Project. She has collected $300 more from Company A than from Company B and $700 more from Company C than from Company A. So far, she has collected $4,300. How much did Company C give?

14. For his birthday, Torin got $50.00 more from his grandmother than from his uncle. His uncle gave him $10.00 less than his cousin. Torin received $135.00 in total. How much did he receive from his cousin?

15. Cassidy loves black and yellow jelly beans. She noticed when she was counting them that she had 7 less than three times as many black jelly beans as she had yellow jelly beans. In total, she counted 225 jelly beans. How many black jelly beans did she have?

16. Mrs. Vargus planted a garden with red and white rose bushes. Because she was studying to be a botanist, she counted the number of blossoms on each bush. She counted 4 times as many red blossoms as white blossoms. In total, she counted 1,420 blossoms. How many red blossoms did she count?

5.3 Consecutive Integer Problems

Consecutive integers follow each other in order

Examples:
1, 2, 3, 4
−3, −4, −5, −6

Algebraic notation:
$n, n+1, n+2, n+3$

Consecutive **even** integers:

2, 4, 6, 8, 10
−12, −14, −16, −18

$n, n+2, n+4, n+6$

Consecutive **odd** integers:

3, 5, 7, 9
−5, −7, −9, −11

$n, n+2, n+4, n+6$

Example 5: The sum of three consecutive odd integers is 63. Find the integers.

Step 1: Represent the three odd integers:
Let n = the first odd integer
$n + 2$ = the second odd integer
$n + 4$ = the third odd integer

Step 2: The sum of the integers is 63, so the algebraic equation is
$n + n + 2 + n + 4 = 63$. Solve for n.
$n = 19$

Solution: the first odd integer = 19
the second odd integer = 21
the third odd integer = 23

Check: Does $19 + 21 + 23 = 63$? Yes, it does.

Example 6: Find three consecutive odd integers such that the sum of the first and second is three less than the third.

Step 1: Represent the three odd integers just like above:
Let n = the first odd integer
$n + 2$ = the second odd integer
$n + 4$ = the third odd integer

Step 2: In this problem, the sum of the first and second integers is three less than the third integer, so the algebraic equation is written as follows:
$n + n + 2 = n + 4 - 3$
$n = -1$

Solution: the first odd integer = −1
the second odd integer = 1
the third odd integer = 3

Check: Is the sum of −1 and 1 three less than 3?
$-1 + 1 = 3 - 3$ or $0 = 0$ Yes, it is.

Solve the following problems.

1. Find three consecutive even integers whose sum is 120.
2. Find three consecutive integers whose sum is -30.
3. The sum of three consecutive odd integers is 51. What are the numbers?
4. Find two consecutive odd integers such that five times the first equals three times the second.
5. Find two consecutive even integers such that seven times the first equals six times the second.
6. Find two consecutive odd numbers whose sum is eighty.

5.4 Inequality Word Problems

Inequality word problems involve staying under a limit or having a minimum goal one must meet.

Example 7: A contestant on a popular game show must earn a minimum of 800 points by answering a series of questions worth 40 points each per category in order to win the game. The contestant will answer questions from each of four categories. Her results for the first three categories are as follows: 160 points, 200 points, and 240 points. Write an inequality which describes how many points, (p), the contestant will need on the last category in order to win.

Step 1: Add to find out how many points she already has. $160 + 200 + 240 = 600$

Step 2: Subtract the points she already has from the minimum points she needs. $800 - 600 = 200$. She must get at least 200 points in the last category to win. If she gets more than 200 points, that is okay, too. To express the number of points she needs, use the following inequality statement:
$p \geq 200$ The points she needs must be greater than or equal to 200.

Solve each of the following problems using inequalities.

1. Stella wants to place her money in a high interest money market account. However, she needs at least $1,000 to open an account. Each month, she set aside some of her earnings in a savings account. In January through June, she added the following amounts to her savings: $121, $206, $138, $212, $109, and $134. Write an inequality which describes the amount of money she can set aside in July to qualify for the money market account.

2. A high school band program will receive $2,000.00 for selling $10,000.00 worth of coupon books. Six band classes participate in the sales drive. Classes 1–5 collect the following amounts of money: $1,400, $2,600, $1,800, $2,450, and $1,550. Write an inequality which describes the amount of money the sixth class must collect so that the band will receive $2,000.

3. A small elevator has a maximum capacity of 1,000 pounds before the cable holding it in place snaps. Six people get on the elevator. Five of their weights follow: 146, 180, 130, 262, and 135. Write an inequality which describes the amount the sixth person can weigh without snapping the cable.

4. A small high school class of 9 students were told they would receive a pizza party if their class average was 92% or higher on the next exam. Students 1–8 scored the following on the exam: 86, 91, 98, 83, 97, 89, 99, and 96. Write an inequality which describes the score the ninth student must make for the class to qualify for the pizza party.

Chapter 5   Algebra Word Problems

Chapter 5 Review

Solve each of the following problems.

1. Deanna is four more than three times older than Ted. The sum of their ages is 60. How old is Ted?

2. Find three consecutive odd numbers whose sum is three hundred and three.

3. The perimeter of a triangle is 48 inches. The second side is four inches longer than the first side. The third side is one inch longer than the second. Find the length of each side.

4. Joe, Craig, and Dylan have a combined weight of 326 pounds. Craig weighs 40 pounds more than Joe. Dylan weighs 12 pounds more than Craig. How many pounds does Craig weigh?

5. Jim takes great pride in decorating his float for the homecoming parade for his high school. With the $5,000 he has to spend, Jim buys 5,000 carnations at $0.30 each, 4,000 tulips at $0.60 each, and 300 irises at $0.25 each. Write an inequality which describes how many roses, r, Jim can buy if roses cost $0.80 each.

6. Mr. Chan wants to sell some or all of his shares of stock in a company. He purchased the 90 shares for $0.50 last month, and the shares are now worth $3.80 each. Write an inequality which describes how much profit, p, Mr. Chan can make by selling his shares.

7. Three consecutive integers have the sum of 54. Find the integers.

8. Lena and Jodie are sisters and together they have 56 bottles of nail polish. Lena bought 4 more than half the bottles. How many did Jodie buy?

Chapter 5 Test

1. Ross is five years older than twice his sister Holly's age. The difference is their ages is 14 years. How old is Holly?

   (A) 9
   (B) 23
   (C) 3
   (D) 18

2. The sum of two numbers is 27. The larger number is 6 more than twice the smaller number. What are the numbers?

   (A) 11, 16
   (B) 19, 8
   (C) 7, 20
   (D) 3, 24

3. The perimeter of a rectangle is 292 feet. The length of the rectangle is 4 feet less than 5 times the width. What is the length and width of the rectangle?

   (A) length = 121, width = 25
   (B) length = 114.3, width = 23.7
   (C) length = 25, width = 121
   (D) length = 121.7, width = 24.3

4. Jesse and Larry entered a pie eating contest. Jesse ate 5 less than twice as many pies as Larry. They ate a total of 16 pies. How many pies did Larry eat?

   (A) 3.7 pies
   (B) 9 pies
   (C) 21 pies
   (D) 7 pies

5. Janet and Artie want to play tug-of-war. Artie pulls with 200 pounds of force while Janet pulls with 60 pounds of force. In order to make this a fair contest, Janet enlists the help of her friends Trudi, Sherri, and Bridget who pull with 20, 25, and 55 pounds respectively. Write an inequality describing the minimum amount Janet's fourth friend, Tommy, must pull to beat Artie.

   (A) $x > 40$ pounds of force
   (B) $x < 40$ pounds of force
   (C) $x > 100$ pounds of force
   (D) $x < 100$ pounds of force

6. There is a new bike that Bianca has had her eye on for a few weeks. The bike costs $75. Her allowance is 10 dollars per week. If she saves 60% of her allowance each week, write an inequality that describes the minimum amount of weeks, y, that Bianca must save in order to buy that bike.

   (A) $y > 75 - 0.6\,(10)$

   (B) $y > 45\,(10)$

   (C) $y > \dfrac{75}{0.6\,(10)}$

   (D) $10 > \dfrac{75}{0.6y}$

Chapter 6
Polynomials

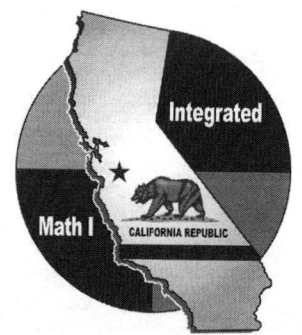

This chapter covers the following CA Integrated Math I standards:

| Algebra | 10.0 |
| | 13.0 |

Polynomials are algebraic expressions which include **monomials** containing one term, **binomials** which contain two terms, and **trinomials**, which contain three terms. Expressions with more than three terms are called **polynomials**. **Terms** are separated by plus and minus signs.

EXAMPLES

Monomials	Binomials	Trinomials	Polynomials
$5f$	$5t + 20$	$x^2 + 4x + 3$	$x^3 - 3x^4 + 3x - 20$
$3x^3$	$20 - 8g$	$7x^4 - 6x - 2$	$p^5 + 4p^3 + p^4 + 20p - 7$
$5g^4$	$7x^4 + 8x$	$y^5 + 27y^4 + 200$	
4	$6x^3 - 9x$		

6.1 Adding and Subtracting Monomials

Two **monomials** are added or subtracted as long as the **variable and its exponent** are the **same**. This is called combining like terms. Use the same rules you used for adding and subtracting integers

Example 1: $\quad 5x + 7x = 12x \quad \begin{array}{r} 3x^5 \\ -9x^5 \\ \hline -6x^5 \end{array} \quad 4x^4 - 20x^4 = -16x^4 \quad \begin{array}{r} 7y \\ +4y \\ \hline 11y \end{array} \quad 6y^3 - 7y^3 = -y^3$

Remember: When the integer in front of the variable is "1", the one is usually not written. $1x^4$ is the same as x^4, and $-1x$ is the same as $-x$.

Add or subtract the following monomials.

1. $4x^4 + 7x^4 =$
2. $7t + 9t =$
3. $20y^3 - 4y^3 =$
4. $6g - 9g =$
5. $8y^4 + 9y^4 =$

6. $s^7 + s^7 =$
7. $-4x - 5x =$
8. $5w^4 - w^4 =$
9. $z^5 + 20z^5 =$
10. $-k + 4k =$

11. $3x^4 - 7x^4 =$
12. $20t + 4t =$
13. $-8v^3 + 20v^3 =$
14. $-4x^3 + x^3 =$
15. $20y^5 - 7y^5 =$

16. y^5
    $+4y^5$

18. $9t^4$
    $+8t^4$

20. $7w^4$
    $+9w^4$

22. $-7z$
    $+20z$

24. $8t^3$
    $-6t^3$

17. $5x^3$
    $-20x^3$

19. $-4y$
    $-5y$

21. $22t^3$
    $-5t^3$

23. $5w^7$
    $+w^7$

25. $3x$
    $+9x$

6.2 Adding Polynomials

When adding **polynomials,** make sure the exponents and variables are the same on the terms you are combining. The easiest way is to put the terms in columns with **like exponents** under each other. Each column is added as a separate problem. Fill in the blank spots with zeros if it helps you keep the columns straight. You never carry to the next column when adding polynomials.

Example 2: Add $3x^4 + 25$ and $7x^4 + 4x$

$$\begin{array}{r} 3x^4 + 0x + 25 \\ (+)\ 7x^4 + 4x + 0 \\ \hline 10x^4 + 4x + 25 \end{array}$$

Example 3: $(5x^3 - 4x) + (-x^3 - 5)$

$$\begin{array}{r} 5x^3 - 4x + 0 \\ (+) - x^3 + 0x - 5 \\ \hline 4x^3 - 4x - 5 \end{array}$$

Add the following polynomials.

1. $y^4 + 3y + 4$ and $4y^4 + 5$
2. $(7y^4 + 5y - 6) + (4y^4 - 7y + 9)$
3. $-4x^4 + 7x^3 + 5x - 2$ and $3x^4 - x + 4$
4. $-p + 5$ and $7p^4 - 4p + 4$
5. $(w - 4) + (w^4 + 4)$
6. $5t^4 - 7t - 8$ and $9t + 4$
7. $t^5 + t + 9$ and $4t^3 + 5t - 5$
8. $(s^4 + 3s^3 - 4) + (-4s^3 + 5)$
9. $(-v^4 + 8v - 9) + (5v^3 - 6v + 5)$
10. $6m^4 - 4m + 20$ and $m^4 - m - 9$
11. $-x + 5$ and $3x^4 + x - 4$
12. $(9t^4 + 3t) + (-8t^4 - t + 5)$
13. $(3p^5 + 4p^4 - 2) + (-7p^4 - p + 9)$
14. $20s^4 + 24s^3 + 4s$ and $s^4 + s^3 + s$
15. $(-20b^4 + 8b + 4) + (-b^4 + 6b + 20)$
16. $27c^4 - 22c + 7$ and $-8c^4 + 3c - 20$
17. $4c^4 + 7c^3 + 3$ and $5c^4 + 4c^3 + 2$
18. $3x^4 + -25x^3 + 27$ and $8x^3 - 24$
19. $(-x^4 + 4x - 5) + (3x^4 - 3)$
20. $(y^4 - 22y + 20) + (-23y^4 + 7y - 5)$
21. $3d^7 - 5d^3 + 8$ and $4d^5 - 4d^3 - 4$
22. $(6t^7 - t^3 + 28) + (5t^7 + 8t^3)$
23. $5p^4 - 9p + 20$ and $-p^4 - 3p - 7$
24. $40b^3 + 27b$ and $-5b^4 - 7b + 25$
25. $(-4w + 22) + (w^3 + w - 5)$
26. $(47z^4 + 23z + 9) + (z^4 - 4z - 20)$

6.3 Subtracting Polynomials

When you subtract polynomials, it is important to remember to change all the signs in the subtracted polynomial (the subtrahend) and then add.

Example 4: $(5y^4 + 9y + 20) - (4y^4 + 6y - 5)$

Step 1: Copy the subtraction problem into vertical form. Make sure you line up the terms with like exponents under each other just like you did for adding polynomials.

$$\begin{array}{r} 5y^4 + 9y + 20 \\ (-)\ 4y^4 + 6y - 5 \\ \hline \end{array}$$

Step 2: Change the subtraction sign to addition and all the signs of the subtracted polynomial to the opposite sign. The bottom polynomial in the problem becomes $-4y^4 - 6y + 5$.

Step 3: Add:
$$\begin{array}{r} 5y^4 + 9y + 20 \\ (+)\ -4y^4 - 6y + 5 \\ \hline y^4 + 3y + 25 \end{array}$$

Subtract the following polynomials.

1. $(4x^4 + 7x + 4) - (x^4 + 3x + 2)$
2. $(9y - 5) - (5y + 3)$
3. $(-5t^4 + 22t^3 + 3) - (5t^4 - t^3 - 7)$
4. $(-3w^4 + 20w - 7) - (-7w^4 - 7)$
5. $(6a^7 - a^3 + a) - (8a^7 + a^4 - 3a)$
6. $(25c^5 + 40c^4 + 20) - (8c^5 + 7c^4 + 24)$
7. $(7x^4 - 20x) - (-8x^4 + 5x + 9)$
8. $(-9y^4 + 24y^3 - 20) - (3y^3 + y + 20)$
9. $(-3h^4 - 8h + 8) - (7h^4 + 5h + 20)$
10. $(20k^3 - 9) - (k^4 - 5k^3 + 7)$
11. $(x^4 - 7x + 20) - (6x^4 - 7x + 8)$
12. $(24p^4 + 5p) - (20p - 4)$
13. $(-4m - 9) - (6m + 4)$
14. $(4y^4 + 23y^3 - 9y) - (5y^4 + 4y^3 - 8y)$
15. $(8g + 3) - (g^4 + 5g - 7)$
16. $(-9w^3 + 5w) - (-5w^4 - 20w^3 - w)$
17. $(x^4 + 24x^3 - 20) - (4x^4 + 3x^3 + 2)$
18. $(4a^4 + 4a + 4) - (-a^4 + 3a + 3)$
19. $(c + 220) - (3c^4 - 8c + 4)$
20. $(-6v^4 + 24v) - (3v^4 + 4v + 6)$
21. $(3b^4 + 5b^3 + 7) - (8b^3 - 9)$
22. $(7x^4 + 27x^3 - 5) - (-5x^4 + 5x^3)$
23. $(9y^4 - 4) - (22y^4 - 4y - 3)$
24. $(-z^4 - 7z - 9) - (3z^4 - 7z + 7)$

6.3 Subtracting Polynomials

A subtraction of polynomials problem may be stated in sentence form. Study the examples below.

Example 5: Subtract $-7x^3 + 5x - 3$ from $3x^3 + 5x^4 - 6x$.

Step 1: Copy the problem in columns with terms with the same exponent and variable under each other. Notice the second polynomial in the sentence will be the top polynomial of the problem.

$$\begin{array}{r} 3x^3 + 5x^4 - 6x \\ (-) -7x^3 + 5x - 3 \\ \hline \end{array}$$

Since this is a subtraction problem, change all the signs of the terms in the bottom polynomial. Then add.

$$\begin{array}{r} 3x^3 + 5x^4 - 6x \\ (+) 7x^3 - 5x + 3 \\ \hline 10x^3 + 5x^4 - 11x + 3 \end{array}$$

Example 6: From $6y^4 + 4$ subtract $5y^4 - 3y + 9$

In a problem phrased like this one, the first polynomial will be on top, and the second will be on bottom. Change the signs on the bottom polynomial and then add.

$$\begin{array}{r} 6y^4 + 4 \\ (-) 5y^4 - 3y + 9 \\ \hline \end{array} \longrightarrow \begin{array}{r} 6y^4 + 4 \\ (+) -5y^4 + 3y - 9 \\ \hline y^4 + 3y - 5 \end{array}$$

Solve the following subtraction problems.

1. Subtract $3x^4 + 4x - 7$ from $7x^4 + 4$
2. From $7y^3 - 6y + 20$ subtract $9y^3 - 20$
3. From $5m^4 - 5m + 8$ subtract $4m - 3$
4. Subtract $9z^4 + 3z + 4$ from $5z^4 - 8z + 9$
5. Subtract $t^4 + 20t^3 - 7$ from $-t^4 - 4t^3 - 7$
6. Subtract $-8b^3 - 4b + 5$ from $-b^4 + b + 6$
7. From $20y^3 + 40$ subtract $7y^3 - 7$
8. From $25t^4 - 6t - 9$ subtract $5t^4 - 3t + 4$
9. Subtract $3p^4 + p - 4$ from $-8p^4 - 7p + 4$
10. Subtract $x^3 + 9$ from $-4x^4 + 3x^3 + 20$
11. Subtract $24a^4 + 20$ from $-a^4 + a^3 - 2$
12. From $6m^4 + 3m + 2$ subtract $-6m^4 - 3m$
13. From $-3z^4 - 23z^3 - 4$ subtract $-40z^3 + 40$
14. Subtract $20c^4 + 20$ from $9c^4 - 7c + 3$
15. Subtract $b^4 + b - 7$ from $7b^4 - 5b + 7$
16. Subtract $-3x - 5$ from $3x^4 + x + 20$
17. From $27y^4 + 4$ subtract $5y^4 + 3y + 8$
18. Subtract $3g^4 - 7g + 7$ from $20g^4 - 3g - 5$
19. From $-8m^4 - 9m$ subtract $3m^4 + 8$
20. Subtract $x + 2$ from $7x + 7$
21. Subtract $c^4 + c + 4$ from $-c^4 - c - 4$
22. From $6t^4 + 9t^3 - 5t + 4$ subtract $t^3 + 3t$

Chapter 6   Polynomials

6.4   Multiplying Monomials

When two monomials have the **same variable**, you can multiply them. Then, add the **exponents** together. If the variable has no exponent, it is understood that the exponent is 1.

Example 7: $5x^5 \times 3x^4 = 15x^9$ $\qquad$ $4y \times 7y^4 = 28y^5$

Multiply the following monomials.

1. $6a \times 20a^7$
2. $4x^6 \times 7x^3$
3. $5y^3 \times 3y^4$
4. $20t^4 \times 4t^4$
5. $4p^7 \times 5p^4$
6. $20b^4 \times 9b$
7. $3c^3 \times 3c^3$

8. $4d^9 \times 20d^4$
9. $6k^3 \times 7k^4$
10. $8m^7 \times m$
11. $22z \times 4z^8$
12. $3w^5 \times 6w^7$
13. $5x^5 \times 7x^3$
14. $7n^4 \times 3n^3$

15. $9w^8 \times w$
16. $20s^6 \times 7s^3$
17. $5d^7 \times 5d^7$
18. $7y^4 \times 9y^6$
19. $8t^{20} \times 3t^7$
20. $6p^9 \times 4p^3$
21. $x^3 \times 4x^3$

When problems include negative signs, follow the rules for multiplying integers.

22. $-8s^5 \times 7s^3$
23. $-6a \times -20a^7$
24. $5x \times -x$
25. $-3y^4 \times -y^3$
26. $-7b^4 \times 3b^7$
27. $20c^5 \times -4c$
28. $-5t^3 \times 9t^3$

29. $20d \times -9d^8$
30. $-3g^6 \times -4g^3$
31. $-8s^5 \times 8s^3$
32. $-d^3 \times -4d$
33. $22p \times -4p^7$
34. $-7x^8 \times -3x^3$
35. $9z^5 \times 8z^5$

36. $-5w \times -7w^9$
37. $-7y^5 \times 6y^4$
38. $20x^3 \times -8x^7$
39. $-a^5 \times -a$
40. $-8k^4 \times 3k$
41. $-27t^4 \times -t^5$
42. $3x^9 \times 20x^4$

6.5 Multiplying Monomials with Different Variables

Warning: You cannot add the exponents of variables that are different.

Example 8: $(-5wx)(6w^3x^4)$

To work this problem, first multiply the whole numbers: $-5 \times 6 = -30$. Then multiply the w's: $w \times w^3 = w^4$. Last, multiply the x's: $x \times x^4 = x^5$. The answer is $-30w^4x^5$.

Multiply the following monomials.

1. $(4x^4y^4)(-5xy^3) =$
2. $(20p^3q^5)(4p^4q) =$
3. $(-3t^5v^4)(t^4v) =$
4. $(8w^3z^4)(3wz) =$
5. $(-4st^6)(-9s^4t) =$
6. $(xy^3)(5x^4y^4) =$
7. $(7y^4z)(3y^5z^4) =$
8. $(-3a^4b^4)(-5ab^3) =$
9. $(-7c^3d^4)(4c^5d^7) =$
10. $(20x^5y^4)(3x^3y) =$
11. $(6f^3g^7)(-f^3g) =$
12. $(-5a^3v^5)(9a^5v) =$
13. $(7m^9n^7)(8m^4n^5) =$
14. $(8w^7y^3)(3wy) =$
15. $(4x^5z^4)(-20x^4z^5) =$
16. $(-5a^8c^{20})(4a^4c) =$
17. $(-bd^6)(-b^4d) =$
18. $(3x^5y^4)(20x^3y^3) =$
19. $(20p^4y)(7p^7y^3) =$
20. $(-4a^8x^4)(6ax^4) =$
21. $(9c^5d^3)(-4c^4d^4) =$

Multiplying three monomials works the same way. The first one is done for you.

22. $(3st)(5s^3t^4)(4s^4t^5) = 60s^8t^{10}$
23. $(xy)(x^4y^4)(4x^3y^4) =$
24. $(4a^4b^4)(a^3b^3)(4ab) =$
25. $(5y^4z^5)(4y^3)(4z^4) =$
26. $(7cd^3)(3c^4d^4)(d^4) =$
27. $(4w^4x^3)(3x^5)(4w^3) =$
28. $(a^5d^4)(ad)(a^4d^3) =$
29. $(8x^3t)(4t^5)(x^4t^4) =$
30. $(p^4y^4)(5py)(p^3y^3) =$
31. $(5x^3y)(7xy^3)(4y^5) =$
32. $(9xy^4)(x^4y^3)(4x^4y) =$
33. $(6p^3t)(4t^3)(p^4) =$
34. $(3bc)(b^4c)(5c^3) =$
35. $(4y^5z^7)(4y^6)(y^4z^4) =$
36. $(5p^3r^3)(5r^4)(p^4r) =$
37. $(a^5z^6)(6a^4z^4)(3z^3) =$
38. $(7c^3)(6d^4)(4c^4d) =$
39. $(20s^8t^4)(3st)(s^4t^3) =$
40. $(3a^3b^5)(4b^3)(3a^4) =$
41. $(7wz)(w^3z^3)(3w^4z^3) =$

6.6 Dividing Monomials

When simplifying monomial fractions with exponents, all exponents need to be positive. If there are negative exponents in your answer, put the base with its negative exponent below the fraction line and remove the negative sign. Two variables that are alike should not appear in both the denominator and numerator of a reduced expression. If you have the same variable in both the denominator and numerator of the fraction, the expression has not been fully reduced.

Example 9: $\dfrac{55x^4y^4}{11x^6y^6}$

Step 1: Reduce the whole numbers first. $\dfrac{55}{11} = 5$

Step 2: Simplify the x's. $\dfrac{x^4}{x^6} = x^{4-6} = x^{-2} = \dfrac{1}{x^2}$

Step 3: Simplify the y's. $\dfrac{y^4}{y^6} = y^{4-6} = y^{-2} = \dfrac{1}{y^2}$

Therefore $\dfrac{55x^4y^4}{22x^6y^6} = \dfrac{5}{x^2y^2}$

Simplify the expressions below. All answers should only have positive exponents.

1. $\dfrac{7xy^3}{x(4x^3)y^5}$

2. $\dfrac{4a^4b^7}{3a^5b^4}$

3. $\dfrac{8(4a^4)b^5}{9ab^6}$

4. $\dfrac{26(x^4y^5)^3}{40xy}$

5. $\dfrac{20a^5b^4}{7a^6b^3}$

6. $\dfrac{7(9x^4y^3)}{5(x^4y^4)^4}$

7. $\dfrac{24(3a^4)b^4}{6a^4b^4}$

8. $\dfrac{(6x^3y^5)^4}{(4x^7y)^3}$

9. $\dfrac{27a^4b^3}{3a^7b^6}$

10. $\dfrac{33x^7y^3}{44x^8y^7}$

11. $\dfrac{27(4a^6b^8)}{42a^3b^6}$

12. $\dfrac{30x^5y^4}{6(x^4y^4)^4}$

13. $\dfrac{20(9ab^7)}{40a^4b^4}$

14. $\dfrac{20x^{20}y^8}{57x^7y^3}$

15. $\dfrac{(a^5b^8)^5}{a^9b^8}$

16. $\dfrac{8(x^3y^7)}{9x^4y^5}$

17. $\dfrac{20(a^3b^5)}{5a^7b^4}$

18. $\dfrac{48(x^4y^7)^4}{42x^3y^5}$

6.7 Multiplying Monomials by Polynomials

In the chapter on solving multi-step equations, you learned to remove parentheses by multiplying the number outside the parentheses by each term inside the parentheses: $4(5x - 8) = 9x - 25$. Multiplying monomials by polynomials works the same way.

Example 10: $-7t(4t^4 - 8t + 20)$

Step 1: Multiply $-7t \times 4t^4 = \mathbf{-28t^5}$

Step 2: Multiply $-7t \times -8t = \mathbf{56t^2}$

Step 3: Multiply $-7t \times 20 = \mathbf{-140t}$

Step 4: Arrange the answers horizontally in order: $\mathbf{-28t^5 + 56t^2 - 140t}$

Remove parentheses in the following problems.

1. $3x(3x^4 + 5x - 2)$
2. $5y(y^3 - 8)$
3. $8a^4(4a^4 + 3a + 4)$
4. $-7d^3(d^4 - 7d)$
5. $4w(-5w^4 + 3w - 9)$
6. $9p(p^3 - 6p + 7)$
7. $-20b^4(-4b + 7)$
8. $4t(t^4 - 5t - 20)$
9. $20c(5c^4 + 3c - 8)$
10. $6z(4z^5 - 7z^4 - 5)$
11. $-20t^4(3t^4 + 7t + 6)$
12. $c(-3c - 7)$
13. $3p(-p^4 + p^3 - 20)$
14. $-k^4(4k + 5)$
15. $-3(5m^4 - 7m + 9)$
16. $6x(-8x^3 + 20)$
17. $-w(w^4 - 5w + 8)$
18. $4y(7y^4 - y)$
19. $3d(d^7 - 8d^3 + 5)$
20. $-7t(-5t^4 - 9t + 2)$
21. $8(4w^4 - 20w + 5)$
22. $3y^4(y^4 - 22)$
23. $v^4(v^4 + 3v + 3)$
24. $9x(4x^3 + 3x + 2)$
25. $-7d(5d^4 + 8d - 4)$
26. $-k^4(-3k + 6)$
27. $3x(-x^4 - 7x + 7)$
28. $5z(5z^5 - z - 8)$
29. $-7y(20y^3 - 3)$
30. $4b^4(8b^4 + 5b + 5)$

Chapter 6 Polynomials

6.8 Dividing Polynomials by Monomials

Example 11: $$\frac{-8wx + 6x^2 - 16wx^2}{2wx}$$

Step 1: Rewrite the problem. Divide each term from the top by the denominator, $2wx$.

$$\frac{-8wx}{2wx} + \frac{6x^2}{2wx} + \frac{-16wx^2}{2wx}$$

Step 2: Simplify each term in the problem. Then combine like terms.

$$-4 + \frac{3x}{w} - 8x$$

Simplify each of the following.

1. $\dfrac{bc^4 - 9bc - 4b^4c^4}{4bc}$

2. $\dfrac{3jk^4 + 24k + 20j^4k}{3jk}$

3. $\dfrac{7x^4y - 9xy^4 + 4y^3}{4xy}$

4. $\dfrac{26st^4 + st - 24s}{5st}$

5. $\dfrac{5wx^4 + 6wx - 24w^3}{4wx}$

6. $\dfrac{cd^4 + 20cd^3 + 26c^4}{4cd}$

7. $\dfrac{y^4z^3 - 4yz - 9z^4}{-4yz^4}$

8. $\dfrac{a^4b + 4ab^4 - 25ab^3}{4a^4}$

9. $\dfrac{pr^4 + 6pr + 9p^4r^4}{4pr^4}$

10. $\dfrac{6xy^4 - 3xy + 29x^4}{-3xy}$

11. $\dfrac{6x^4y + 24xy - 45y^4}{6xy}$

12. $\dfrac{7m^4n - 20mn - 47n^4}{7mn}$

13. $\dfrac{st^4 - 20st - 26s^4t^4}{4st}$

14. $\dfrac{8jk^4 - 25jk - 63j^4}{8jk}$

6.9 Removing Parentheses and Simplifying

In the following problem, you must multiply each set of parentheses by the numbers and variables outside the parentheses, and then add the polynomials to simplify the expressions.

Example 12: $9x(4x^4 - 7x + 8) - 3x(5x^4 + 3x - 9)$

Step 1: Multiply to remove the first set of parentheses.

$9x(4x^4 - 7x + 8) = 36x^5 - 63x^2 + 72x$

Step 2: Multiply to remove the second set of parentheses.

$-3x(5x^4 + 3x - 9) = -15x^5 - 9x^2 + 27x$

Step 3: Copy each polynomial in columns, making sure the terms with the same variable and exponent are under each other. Add to simplify.

$$\begin{array}{r} 36x^5 - 63x^2 + 72x \\ (+) -15x^5 - 9x^2 + 27x \\ \hline 21x^5 - 72x^2 + 99x \end{array}$$

Remove the parentheses and simplify the following problems.

1. $5t(t+8) + 7t(4t^4 - 5t + 2)$

2. $-7y(3y^4 - 7y + 3) - 6y(y^4 - 5y - 5)$

3. $-3(3x^4 + 5x) + 7x(x^4 + 3x + 4)$

4. $4b(7b^4 - 9b - 2) - 3b(5b + 3)$

5. $9d^4(3d + 5) - 8d(3d^4 + 5d + 7)$

6. $7a(3a^4 + 3a + 2) - (-4a^4 + 7a - 5)$

7. $3m(m+8) + 9(5m^4 + m + 5)$

8. $5c^4(-6c^4 - 3c + 4) - 8c(7c^3 + 4c)$

9. $-9w(-w+2) - 5w(3w - 7)$

10. $6p(4p^4 - 5p - 6) + 3p(p^4 + 6p + 20)$

6.10 Multiplying Two Binomials

When you multiply two binomials such as $(x+6)(x-7)$, you must multiply each term in the first binomial by each term in the second binomial. The easiest way is to use the **FOIL** method. If you can remember the word **FOIL**, it can help you keep order when you multiply. The "F" stands for **first**, "O" stands for **outside**, "I" stands for **inside**, and "L" stands for **last**.

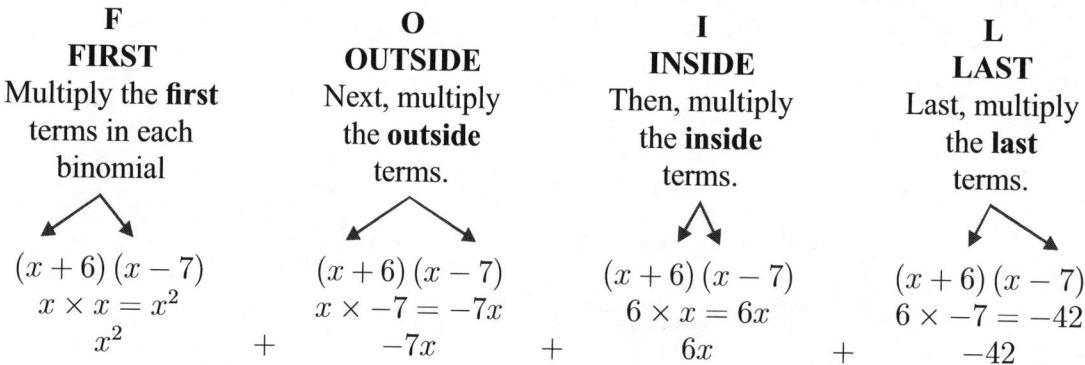

Now just combine like terms, $6x - 7x = -x$, and write your answer.
$(x+6)(x-7) = x^2 - x - 42$.

Note: It is customary for mathematicians to write polynomials in descending order. That means that the term with the highest-number exponent comes first in a polynomial. The next highest exponent is second and so on. When you use the **FOIL** method, the terms will always be in the customary order. You just need to combine like terms and write your answer.

Multiply the following binomials.

1. $(y-8)(y+3)$
2. $(4x+5)(x+20)$
3. $(5b-3)(3b-5)$
4. $(6g+4)(g-20)$
5. $(8k-7)(-5k-3)$
6. $(9v-4)(3v+5)$
7. $(20p+4)(5p+3)$
8. $(3h-20)(-4h-7)$
9. $(w-5)(w-8)$
10. $(6x+2)(x-4)$
11. $(7t+3)(4t-2)$
12. $(5y-20)(5y+20)$
13. $(a+6)(3a+7)$
14. $(3z-9)(z-5)$
15. $(7c+4)(6c+7)$

6.10 Multiplying Two Binomials

16. $(y+3)(y-3)$

17. $(4w-7)(5w+6)$

18. $(8x+2)(x-5)$

19. $(6t-20)(5t-5)$

20. $(7b+6)(6b+4)$

21. $(4z+2)(20z+5)$

22. $(22w-9)(w+3)$

23. $(7d-20)(20d+20)$

24. $(20g+4)(g-4)$

25. $(5p+8)(4p+3)$

26. $(m+7)(m-7)$

27. $(9b-9)(4b-2)$

28. $(z+3)(3z+7)$

29. $(8y-7)(y-3)$

30. $(20x+7)(3x-2)$

31. $(3t+2)(t+20)$

32. $(4w-20)(9w+8)$

33. $(9s-4)(s+5)$

34. $(5k-2)(9k+20)$

35. $(h+24)(h-4)$

36. $(3x+8)(8x+3)$

37. $(4v-6)(4v+6)$

38. $(4x+9)(4x-3)$

39. $(k-2)(6k+24)$

40. $(3w+22)(4w+4)$

41. $(9y-20)(7y-3)$

42. $(6d+23)(d-2)$

43. $(8h+3)(4h+5)$

44. $(7n+20)(7n-7)$

45. $(6z+7)(z-9)$

46. $(5p+7)(4p-20)$

47. $(b+4)(7b+8)$

48. $(20y-3)(9y-8)$

6.11 Simplifying Expressions with Exponents

Example 13: **Simplify** $(4a + 7)^2$

When you simplify an expression such as $(4a + 7)^2$, it is best to write the expression as two binomials and use FOIL to simplify.
$(4a + 7)^2 = (4a + 7)(4a + 7)$
Using FOIL we have $16a^2 + 28a + 28a + 49 = 16a^2 + 56a + 49$

Example 14: **Simplify** $5(3a + 4)^2$

Using order of operations, we must simplify the exponent first.
$= 5(3a + 4)(3a + 4)$
$= 5(9a^2 + 12a + 12a + 16)$
$= 5(9a^2 + 24a + 16)$ Now multiply by 5.
$= 5(9a^2 + 24a + 16) = 45a^2 + 120a + 80$

Note: It is customary for mathematicians to write polynomials in descending order. That means that the term with the highest number exponent comes first in a polynomial. The next highest exponent is second and so on. When you use the **FOIL** method, the terms will always be in the customary order. You just need to combine like terms, and write your answer.

Multiply the following binomials.

1. $(y + 3)^2$
2. $4(4x + 5)^2$
3. $6(5b - 3)^2$
4. $7(6g + 4)^2$
5. $(-5k - 3)^2$
6. $3(-4h - 7)^2$
7. $-4(9v - 4)^2$
8. $(20p + 4)^2$
9. $6(-2h - 2)^2$
10. $6(w - 8)^2$
11. $4(6x + 2)^2$
12. $(3x + 4)^2$
13. $(7t + 3)^2$
14. $3(5y - 20)^2$
15. $9(a + 6)^2$
16. $5(3z - 9)^2$
17. $3(7c + 4)^2$
18. $5(3t + 20)^2$

Chapter 6 Review

Simplify.

1. $3a^4 + 20a^4$
2. $(8x^4y^5)(20xy^7)$
3. $-6z^4(z+3)$
4. $(5b^4)(7b^3)$
5. $8x^4 - 20x^4$
6. $(7p-5) - (3p+4)$
7. $-7t(3t+20)^2$
8. $(3w^3y^4)(5wy^7)$
9. $3(4g+3)^2$
10. $25d^5 - 20d^5$
11. $(8w-5)(w-9)$
12. $27t^4 + 5t^4$
13. $(8c^5)(20c^4)$
14. $(20x+4)(x+7)$
15. $5y(5y^4 - 20y + 4)$
16. $(9a^5b)(4ab^3)(ab)$
17. $(7w^6)(20w^{20})$
18. $9x^3 + 24x^3$
19. $27p^7 - 22p^7$
20. $(3s^5t^4)(5st^3)$
21. $(5d+20)(4d+8)$
22. $5w(-3w^4 + 8w - 7)$
23. $45z^6 - 20z^6$
24. $-8y^3 - 9y^3$
25. $(8x^5)(8x^7)$
26. $28p^4 + 20p^4$
27. $(a^4v)(4av)(a^3v^6)$
28. $5(6y-7)^2$
29. $(3c^4)(6c^9)$
30. $(5x^7y^3)(4xy^3)$
31. Add $4x^4 + 20x$ and $7x^4 - 9x + 4$
32. $5t(6t^4 + 5t - 6) + 9t(3t+3)$
33. Subtract $y^4 + 5y - 6$ from $3y^4 + 8$
34. $4x(5x^4 + 6x - 3) + 5x(x+3)$
35. $(6t-5) - (6t^4 + t - 4)$
36. $(5x+6) + (8x^4 - 4x + 3)$
37. Subtract $7a - 4$ from $a + 20$
38. $(-4y+5) + (5y-6)$
39. $4t(t+6) - 7t(4t+8)$
40. Add $3c - 5$ and $c^4 - 3c - 4$
41. $4b(b-5) - (b^4 + 4b + 2)$
42. $(6k^4 + 7k) + (k^4 + k + 20)$
43. $(q^4r^3)(3qr^4)(4q^5r)$
44. $(7df)(d^5f^4)(4df)$
45. $(8g^4h^3)(g^3h^6)(6gh^3)$
46. $(9v^4x^3)(3v^6x^4)(4v^5x^5)$
47. $(3n^4m^4)(20n^4m)(n^3m^8)$
48. $(22t^4a^4)(5t^3a^9)(4t^6a)$
49. $\dfrac{24(4a^3)b}{3a^4b^{-4}}$
50. $\dfrac{8(g^3h^3)}{5(g^4h)^{-4}}$
51. $\dfrac{26(m^4n^3)^4}{5(m^4n)^{-4}}$
52. $\dfrac{25p^3q^3}{4p^4q}$
53. $\dfrac{9(e^5h^{-4})^{-4}}{36e^4h^7}$
54. $\dfrac{44x^3y^5}{154(x^{-3}y^8)^4}$

Chapter 6 Test

1. $2x^2 + 5x^2 =$

    (A) $10x^4$
    (B) $7x^4$
    (C) $7x^2$
    (D) $10x^2$

2. $-8m^3 + m^3 =$

    (A) $-8m^6$
    (B) $-8m^9$
    (C) $-9m^6$
    (D) $-7m^3$

3. $(6x^3 + x^2 - 5) + (-3x^3 - 2x^2 + 4) =$

    (A) $3x^3 - x^2 - 1$
    (B) $3x^3 - 3x^2 - 1$
    (C) $3x^3 - 3x^2 - 9$
    (D) $-3x^3 - 3x^2 - 1$

4. $(-7c^2 + 5c + 3) + (-c^2 - 7c + 2) =$

    (A) $-3x^3 - 3x^2 - 1$
    (B) $-8c^2 - 2c + 5$
    (C) $-6c^2 - 12c + 5$
    (D) $-8c^2 - 12c + 5$

5. $(5x^3 - 4x^2 + 5) - (-2x^3 - 3x^2) =$

    (A) $3x^3 + x^2 + 5$
    (B) $3x^3 - 7x^2 + 5$
    (C) $7x^3 - x^2 + 5$
    (D) $7x^3 - 7x^2 + 5$

6. $(-z^3 - 4z^2 - 6) - (3z^3 - 6z + 5) =$

    (A) $-4z^3 - 4z^2 + 6z - 11$
    (B) $-2z^3 - 10z - 1$
    (C) $-4z^3 - 10z^2 - 1$
    (D) $-2z^2 + 2z - 11$

7. $(-7d^5)(-3d^2) =$

    (A) $-21d^7$
    (B) $21d^{10}$
    (C) $21d^7$
    (D) $-21d^{10}$

8. $(-5c^3d)(3c^5d^3)(2cd^4) =$

    (A) $30c^{15}d^8$
    (B) $15c^8d^{12}$
    (C) $-17c^{15}d^{12}$
    (D) $-30c^9d^8$

9. $-11j^2 \times -j^4 =$

    (A) $11j^6$
    (B) $11j^8$
    (C) $-11j^6$
    (D) $-11j^8$

10. $-6m^2(7m^2 + 5m - 6) =$

    (A) $-42m^2 + 30m^3 - 36$
    (B) $-42m^4 - 30m^3 + 36m^2$
    (C) $-13m^4 - m^2 + 36m^2$
    (D) $42m^4 - 30m^3 - 36m^2$

11. $-h^2(-4h + 5) =$

    (A) $-4h^3 - 5h^2$
    (B) $4h^3 - 5h^2$
    (C) $-5h^2 - 5h^2$
    (D) $-5h^3 - 5h^2$

12. $\dfrac{4xy^2 - 6xy + 8x^2y}{2xy} =$

   (A) $2xy - 3 + 4x$
   (B) $2y - 3 + 4xy$
   (C) $2y - 3 + 4x$
   (D) $2xy - 3 + 4x^2$

13. $\dfrac{3cd^3 + 6c^2d - 12cd}{3cd} =$

   (A) $cd + 3c - 4cd$
   (B) $d^2 + 2c - 4cd$
   (C) $cd^2 + 2c - 4cd$
   (D) $d^2 + 2c - 4$

14. $4m(m-5) + 3m(2m^2 - 6m + 4) =$

   (A) $6m^3 - 14m^2 - 8m$
   (B) $-8m^2 - 8m - 1$
   (C) $7m - 14m^2 - 1$
   (D) $10m^2 - 26m - 20$

15. $2h(3h^2 - 5h - 2) + 4h(h^2 + 6h + 8) =$

   (A) $6h^3 + 19h^2 + 28h$
   (B) $-8m^2 - 8m - 1$
   (C) $7m - 14m^2 - 1$
   (D) $10h^3 + 14h^2 + 28h$

16. Multiply the following binomials and simplify.

   $(x-3)(x+3)$

   (A) $x^2 - 3x + 3x - 9$
   (B) $x^2 - 9$
   (C) $x^2 + 9$
   (D) $x^2 + 6x + 9$

17. Multiply the following binomials and simplify.

   $(x+9)(x+1)$

   (A) $x^2 + 10x + 9$
   (B) $x^2 + 10x + 10$
   (C) $x^2 + 9x + 9$
   (D) $x^2 + 9x + x + 9$

18. Multiply the following binomials and simplify.

   $(x-2)^2$

   (A) $x^2 - 4x - 4$
   (B) $x^2 - 2x + 4$
   (C) $x^2 - 2x - 4$
   (D) $x^2 - 4x + 4$

Chapter 7
Factoring

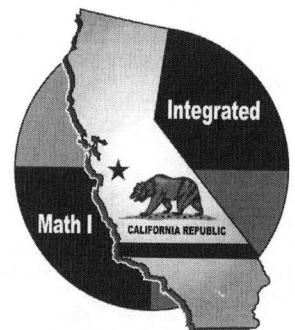

This chapter covers the following CA Integrated Math I standards:

Algebra	11.0
	12.0
	13.0

7.1 Finding the Greatest Common Factor of Polynomials

In a multiplication problem, the numbers multiplied together are called **factors**. The answer to a multiplication problem is a called the **product**.

In the multiplication problem $5 \times 4 = 20$, 5 and 4 are factors and 20 is the product.

If we reverse the problem, $20 = 5 \times 4$, we say we have **factored** 20 into 5×4.

In this chapter, we will factor **polynomials**.

Example 1: Find the greatest common factor of $2y^3 + 6y^2$.

Step 1: Look at the whole numbers. The greatest common factor of 2 and 6 is 2. Factor the 2 out of each term.

$$2\left(y^3 + 3y^2\right)$$

Step 2: Look at the remaining terms, $y^3 + 3y^2$. What are the common factors of each term?

$$\begin{array}{rcl} y^3 & = & y \times \boxed{y \times y} \\ 3y^2 & = & 3 \times \boxed{y \times y} \end{array} \longleftarrow \text{common factors} = y^2$$

Step 2: Factor 2 and y^2 out of each term: $2y^2(y+3)$

Check: $2y^2(y+3) = 2y^3 + 6y^2$

7.1 Finding the Greatest Common Factor of Polynomials

Find the greatest common factor of each of the following.

1. $6x^4 + 18x^2$
2. $14y^3 + 7y$
3. $4b^5 + 12b^3$
4. $10a^3 + 5$
5. $2y^3 + 8y^2$
6. $6x^4 - 12x^2$
7. $18y^2 - 12y$
8. $15a^3 - 25a^2$
9. $4x^3 + 16x^2$
10. $6b^2 + 21b^5$
11. $27m^3 + 18m^4$
12. $100x^4 - 25x^3$
13. $4b^4 - 12b^3$
14. $18c^2 + 24c$
15. $20y^3 + 30y^5$
16. $16x^2 - 24x^5$
17. $15a^4 - 25a^2$
18. $24b^3 + 16b^6$
19. $36y^4 + 9y^2$
20. $42x^3 + 49x$

Factoring larger polynomials with 3 or 4 terms works the same way.

Example 2: $4x^5 + 16x^4 + 12x^3 + 8x^2$

Step 1: Find the greatest common factor of the whole numbers. 4 can be divided evenly into 4, 16, 12, and 8; therefore, 4 is the greatest common factor.

Step 2: Find the greatest common factor of the variables. x^5, x^4, x^3, and x^2 can be divided by x^2, the lowest power of x in each term.

$$4x^5 + 16x^4 + 12x^3 + 8x^2 = 4x^2(x^3 + 4x^2 + 3x + 2)$$

Factor each of the following polynomials.

1. $5a^3 + 15a^2 + 20a$
2. $18y^4 + 6y^3 + 24y^2$
3. $12x^5 + 21x^3 + x^2$
4. $6b^4 + 3b^3 + 15b^2$
5. $14c^3 + 28c^2 + 7c$
6. $15b^4 - 5b^2 + 20b$
7. $t^3 + 3t^2 - 5t$
8. $8a^3 - 4a^2 + 12a$
9. $16b^5 - 12b^4 - 10b^2$
10. $20x^4 + 16x^3 - 24x^2 + 28x$
11. $40b^7 + 30b^5 - 50b^3$
12. $20y^4 - 15y^3 + 30y^2$
13. $4m^5 + 8m^4 + 12m^3 + 6m^2$
14. $16x^5 + 20x^4 - 12x^3 + 24x^2$
15. $18y^4 + 21y^3 - 9y^2$
16. $3n^5 + 9n^3 + 12n^2 + 15n$
17. $4d^6 - 8d^2 + 2d$
18. $10w^2 + 4w + 2$
19. $6t^3 - 3t^2 + 9t$
20. $25p^5 - 10p^3 - 5p^2$
21. $18x^4 + 9x^2 - 36x$
22. $6b^4 - 12b^2 - 6b$
23. $y^3 + 3y^2 - 9y$
24. $10x^5 - 2x^4 + 4x^2$

Chapter 7 Factoring

Example 3: Find the greatest common factor of $4a^3b^2 - 6a^2b^2 + 2a^4b^3$

Step 1: The greatest common factor of the whole numbers is 2.

$$4a^3b^2 - 6a^2b^2 + 2a^4b^3 = 2(2a^3b^2 - 3a^2b^2 + a^4b^3)$$

Step 2: Find the lowest power of each variable that is in each term. Factor them out of each term. The lowest power of a is a^2. The lowest power of b is b^2.

$$4a^3b^2 - 6a^2b^2 + 2a^4b^3 = 2a^2b^2(2a - 3 + a^2b)$$

Factor each of the following polynomials.

1. $3a^2b^2 - 6a^3b^4 + 9a^2b^3$

2. $12x^4y^3 + 18x^3y^4 - 24x^3y^3$

3. $20x^2y - 25x^3y^3$

4. $12x^2y - 20x^2y^2 + 16xy^2$

5. $8a^3b + 12a^2b + 20a^2b^3$

6. $36c^4 + 42c^3 + 24c^2 - 18c$

7. $14m^3n^4 - 28m^3n^2 + 42m^2n^3$

8. $16x^4y^2 - 24x^3y^2 + 12x^2y^2 - 8xy^2$

9. $32c^3d^4 - 56c^2d^3 + 64c^3d^2$

10. $21a^4b^3 + 27a^2b^3 + 15a^3b^2$

11. $4w^3t^2 + 6w^2t - 8wt^2$

12. $5pw^3 - 2p^2q^2 - 9p^3q$

13. $49x^3t^3 + 7xt^2 - 14xt^3$

14. $9cd^4 - 3d^4 - 6c^2d^3$

15. $12a^2b^3 - 14ab + 10ab^2$

16. $25x^4 + 10x - 20x^2$

17. $bx^3 - b^2x^2 + b^3x$

18. $4k^3a^2 + 22ka + 16k^2a^2$

19. $33w^4y^2 - 9w^3y^2 + 24w^2y^2$

20. $18x^3 - 9x^5 + 27x^2$

7.2 Factor By Grouping

Not all polynomials have a common factor in each term. In this case they may sometimes be factored by grouping.

Example 4: Factor $ab + 4a + 2b + 8$

Step 1: Factor an a from the first two terms and a 2 from the last two terms.

$$a(b+4) + 2(b+4)$$

Now the polynomial has two terms, $a(b+4)$ and $2(b+4)$. Notice that $(b+4)$ is a factor of each term.

Step 2: Factor out the common factor of each term:

$$ab + 4a + 2b + 8 = (b+4)(a+2).$$

Check: Multiply using the FOIL method to check.

$$(b+4)(a+2) = ab + 4a + 2b + 8$$

Factor the following polynomials by grouping.

1. $xy + 4x + 2y + 8$
2. $cd + 5c + 4d + 20$
3. $xy - 4x + 6y - 24$
4. $ab + 6a + 3b + 18$
5. $ab + 3a - 5b - 15$
6. $xy - 2x + 6x - 12$
7. $cd + 4c + 4d + 16$
8. $mn - 5m + 3n - 15$
9. $ab + 4a + 3b + 12$
10. $xy + 7x - 4y - 28$
11. $ab - 2a + 8b - 16$
12. $cd + 4c - 5d - 20$
13. $mn + 6m - 2n - 12$
14. $xy - 9x - 3y + 27$
15. $bc - 3b + 5c - 15$
16. $ab + a + 7b + 7$
17. $xy + 4y + 2y + 8$
18. $cd + 9c - d - 9$
19. $ab + 2a - 7b - 14$
20. $xy - 6x - 2y + 12$
21. $wz + 6z - 4w - 24$

Chapter 7 Factoring

7.3 Factoring Trinomials

In the chapter on polynomials, you multiplied binomials (two terms) together, and the answer was a trinomial (three terms).

For example, $(x+6)(x-5) = x^2 + x - 30$

Now, you need to practice factoring a trinomial into two binomials.

Example 5: Factor $x^2 + 6x + 8$

Step 1: When the trinomial is in descending order as in the example above, you need to find a pair of numbers whose sum equals the number in the second term, while their product equals the third term. In the above example, find the pair of numbers that has a sum of 6 and a product of 8.

____ + ____ = 6   and   ____ × ____ = 8

The pair of numbers that satisfy both equations is 4 and 2.

Step 2: Use the pair of numbers in the binomials.

The factors of $x^2 + 6x + 8$ are $(x+4)(x+2)$

Check: To check, use the FOIL method.
$(x+4)(x+2) = x^2 + 4x + 2x + 8 = x^2 + 6x + 8$

Notice, when the second term and the third term of the trinomial are both positive, both numbers in the solution are positive.

Example 6: Factor $x^2 - x - 6$ Find the pair of numbers where:

the sum is -1 and the product is -6

____ + ____ = -1 and   ____ × ____ = -6

The pair of numbers that satisfies both equations is 2 and -3.
The factors of $x^2 - x - 6$ are $(x+2)(x-3)$

Notice, if the second term and the third term are negative, one number in the solution pair is positive, and the other number is negative.

7.3 Factoring Trinomials

Example 7: Factor $x^2 - 7x + 12$ Find the pair of numbers where:
the sum is -7 and the product is 12

_____ + _____ = -7 and   _____ × _____ = 12

The pair of numbers that satisfies both equations is -3 and -4
The factors of $x^2 - 7x + 12$ are $(x-3)(x-4)$.

Notice, if the second term of a trinomial is negative and the third term is positive, both numbers in the solution are negative.

Find the factors of the following trinomials.

1. $x^2 - x - 2$
2. $y^2 + y - 6$
3. $w^2 + 3w - 4$
4. $t^2 + 5t + 6$
5. $x^2 + 2x - 8$
6. $k^2 - 4k + 3$
7. $t^2 + 3t - 10$
8. $x^2 - 3x - 4$
9. $y^2 - 5y + 6$
10. $y^2 + y - 20$
11. $a^2 - a - 6$
12. $b^2 - 4b - 5$
13. $c^2 - 5c - 14$
14. $c^2 - c - 12$
15. $d^2 + d - 6$
16. $x^2 - 3x - 28$
17. $y^2 + 3y - 18$
18. $a^2 - 9a + 20$
19. $b^2 - 2b - 15$
20. $c^2 + 7c - 8$
21. $t^2 - 11t + 30$
22. $w^2 + 13w + 36$
23. $m^2 - 2m - 48$
24. $y^2 + 14y + 49$
25. $x^2 + 7x + 10$
26. $a^2 - 7a + 6$
27. $d^2 - 6d - 27$

Chapter 7   Factoring

7.4   More Factoring Trinomials

Sometimes a trinomial has a greatest common factor which must be factored out first.

Example 8:   Factor $4x^2 + 8x - 32$

Step 1:   Begin by factoring out the greatest common factor, 4.

$4(x^2 + 2x - 8)$

Step 2:   Factor by finding a pair of numbers whose sum is 2 and product is -8.
4 and -2 will work, so

$4(x^2 + 2x - 8) = 4(x+4)(x-2)$

Check: Multiply to check. $4(x+4)(x-2) = 4x^2 + 8x - 32$

Factor the following trinomials. Be sure to factor out the greatest common factor first.

1. $2x^2 + 6x + 4$
2. $3y^2 - 9y + 6$
3. $2a^2 + 2a - 12$
4. $4b^2 + 28b + 40$
5. $3y^2 - 6y - 9$
6. $10x^2 + 10x - 200$
7. $5c^2 - 10c - 40$
8. $6d^2 + 30d - 36$
9. $4x^2 + 8x - 60$
10. $6a^2 - 18a - 24$
11. $5b^2 + 40b + 75$
12. $3c^2 - 6c - 24$
13. $2x^2 - 18x + 28$
14. $4y^2 - 20y + 16$
15. $7a^2 - 7a - 42$
16. $6b^2 - 18b - 60$
17. $11d^2 + 66d + 88$
18. $3x^2 - 24x + 45$

7.5 Factoring More Trinomials

Some trinomials have a whole number in front of the first term that cannot be factored out of the trinomial. The trinomial can still be factored.

Example 9: Factor $2x^2 + 5x - 3$

Step 1: To get a product of $2x^2$, one factor must begin with $2x$ and the other with x.

$(2x \quad)(x \quad)$

Step 2: Now think: What two numbers give a product of -3? The two possibilities are 3 and -1 or -3 and 1. We know they could be in any order so there are 4 possible arrangements.

$(2x + 3)(x - 1)$
$(2x - 3)(x + 1)$
$(2x + 1)(x - 3)$
$(2x - 1)(x + 3)$

Step 3: Multiply each possible answer until you find the arrangement of the numbers that works. Multiply the outside terms and the inside terms and add them together to see which one will equal $5x$.

$(2x + 3)(x - 1) = 2x^2 + x - 3$
$(2x - 3)(x + 1) = 2x^2 - x - 3$
$(2x + 1)(x - 3) = 2x^2 - 5 - 3$
$\boxed{(2x - 1)(x + 3) = 2x^2 + 5x - 3}$ ⟵ This arrangement works, therefore:

The factors of $2x^2 + 5x - 3$ are $(2x - 1)(x + 3)$

Alternative: You can do some of the multiplying in your head. For the above example, ask yourself the following question: What two numbers give a product of -3 and give a sum of 5 (the whole number in the second term) when one number is first multiplied by 2 (the whole number in front of the first term)? The pair of numbers, -1 and 3, have a product of -3 and a sum of 5 when the 3 is first multiplied by 2. Therefore, the 3 will go opposite the factor with the $2x$ so that when the terms are multiplied, you get -5.

You can use this method to at least narrow down the possible pairs of numbers when you have several from which to choose.

Chapter 7   Factoring

Factor the following trinomials.

1. $3y^2 + 14y + 8$
2. $5a^2 + 24a - 5$
3. $7b^2 + 30b + 8$
4. $2c^2 - 9c + 9$
5. $2y^2 - 7y - 15$
6. $3x^2 + 4x + 1$
7. $7y^2 + 13y - 2$
8. $11a^2 + 35a + 6$
9. $5y^2 + 17y - 12$
10. $3a^2 + 4a - 7$
11. $2a^2 + 3a - 20$
12. $5b^2 - 13b - 6$
13. $3y^2 - 17x + 36$
14. $2x^2 - 17x + 36$
15. $11x^2 - 29x - 12$
16. $5c^2 + 2c - 16$
17. $7y^2 - 30y + 27$
18. $2x^2 - 3x - 20$
19. $5b^2 + 24b - 5$
20. $7d^2 + 18d + 8$
21. $3x^2 - 20x + 25$
22. $2a^2 - 7a - 4$
23. $5m^2 + 12m + 4$
24. $9y^2 - 5y - 4$
25. $2b^2 - 13b + 18$
26. $7x^2 + 31x - 20$
27. $3c^2 - 2c - 21$

7.6 Factoring the Difference of Two Squares

The product of a term and itself is called a **perfect square**.

25 is a perfect square because $5 \times 5 = 25$
49 is a perfect square because $7 \times 7 = 49$

Any variable with an even exponent is a perfect square.

y^2 is a perfect square because $y \times y = y^2$
y^4 is a perfect square because $y^2 \times y^2 = y^4$

When two terms that are both perfect squares are subtracted, factoring those terms is very easy. To factor the difference of perfect squares, you use the square root of each term, a plus sign in the first factor, and a minus sign in the second factor.

Example 10: Factor $4x^2 - 9$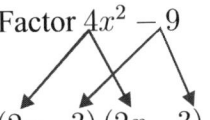

This example has two terms which are both perfect squares, and the terms are subtracted.

Step 1: $(2x \quad 3)(2x \quad 3)$

Find the square root of each term.
Use the square roots in each of the factors.

Step 2: $(2x + 3)(2x - 3)$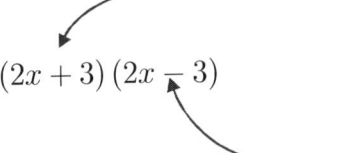

Use a plus sign in one factor and a minus sign in the other factor.

Check: Multiply to check. $(2x + 3)(2x - 3) = 4x^2 - 6x + 6x - 9 = 4x^2 - 9$

The inner and outer terms add to zero.

Example 11: Factor $81y^4 - 1$

Step 1: $(9y^2 + 1)(9y^2 - 1)$

Factor like the example above. Notice, the second factor is also the difference of two perfect squares.

Step 2: $(9y^2 + 1)(3y + 1)(3y - 1)$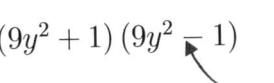

Factor the second term further. **Note: You cannot factor the sum of two perfect squares.**

Check: Multiply in reverse to check your answer.
$(9y^2 + 1)(3y + 1)(3y - 1) = (9y^2 + 1)(9y^2 - 3y + 3y - 1) =$
$(9y^2 + 1)(9y^2 - 1) = 81y^4 + 9y^2 - 9y^2 - 1 = 81y^4 - 1$

Chapter 7 Factoring

Factor the following differences of perfect squares.

1. $64x^2 - 49$
2. $4y^4 - 25$
3. $9a^4 - 4$
4. $25c^4 - 9$
5. $64y^2 - 9$
6. $x^4 - 16$
7. $49x^2 - 4$
8. $4d^2 - 25$
9. $9a^2 - 16$
10. $100y^4 - 49$
11. $c^4 - 36$
12. $36x^2 - 25$
13. $25x^2 - 4$
14. $9x^4 - 64$
15. $49x^2 - 100$
16. $16x^2 - 81$
17. $9y^4 - 1$
18. $49c^2 - 25$
19. $25d^2 - 64$
20. $36a^4 - 49$
21. $16x^4 - 16$
22. $b^2 - 25$
23. $c^4 - 144$
24. $9y^2 - 4$
25. $81x^4 - 16$
26. $4b^2 - 36$
27. $9w^2 - 9$
28. $64a^2 - 25$
29. $49y^2 - 121$
30. $x^6 - 9$

7.7 Simplifying Rational Expressions

We will use what we learned so far in this chapter to factor the terms in the numerator and the denominator when possible, then simplify the rational expression.

Example 12: Simplify $\dfrac{c^2 - 25}{c^2 + 5c}$

Step 1: The numerator is the difference of two perfect squares, so it can be easily factored as in the previous section. Use the square root of each of the terms in the parentheses, with a plus sign in one and a minus sign in the other.
$c^2 - 25 = (c - 5)(c + 5)$

Step 2: Find the greatest common factor in the denominator and factor it out. In this case, it is the variable c.
$c^2 + 5c = c(c + 5)$

Step 3: Simplify $\dfrac{c^2 - 25}{c^2 + 5c} = \dfrac{(c - 5)(c + 5)}{c(c + 5)} = \dfrac{c - 5}{c}$

Simplify the rational expressions. Check for perfect squares and common factors.

1. $\dfrac{25x^2 - 4}{5x^2 - 2x}$

2. $\dfrac{64c^2 - 25}{8c^2 + 5c}$

3. $\dfrac{36a^2 - 49}{6a^2 - 7a}$

4. $\dfrac{x^2 - 9}{x^2 + 3x}$

5. $\dfrac{9a^2 - 16}{3a^2 - 4a}$

6. $\dfrac{16x^2 - 81}{4x^2 - 9x}$

7. $\dfrac{49x^2 - 100}{7x^2 + 10x}$

8. $\dfrac{x^4 - 16}{x^2 + 2x}$

9. $\dfrac{4y^2 - 36}{2y^2 + 6y}$

10. $\dfrac{81y^4 - 16}{9y^2 + 4}$

11. $\dfrac{25x^4 - 225}{5x^2 + 15}$

12. $\dfrac{3y^3 + 9}{y^9 - 9}$

Chapter 7 Factoring

7.8 Adding Rational Expressions

Rational expressions are fractions that can have variables in the numerator and/or the denominator. Adding rational expressions is similar to adding fractions because a rational expression is a fraction.

Example 13: Add: $\dfrac{x-3}{x-2} + \dfrac{5}{x+3}$

Step 1: Just like adding fractions, we must find a common denominator by the finding the least common multiple of the denominators. The least common multiple of $x-2$ and $x+3$ is $(x-2)(x+3)$.

Step 2: Set up the algebra problem like a fraction problem and find the numerators.

$$\dfrac{x-3}{x-2} = \dfrac{(x-3)(x+3)}{(x-2)(x+3)}$$

$$+ \dfrac{5}{x+3} = \dfrac{5x-10}{(x+3)(x-2)}$$

Step 3: Add: $\dfrac{(x-3)(x+3) + 5x - 10}{(x+3)(x-2)}$

Step 4: Simplify: $\dfrac{x^2 - 9 + 5x - 10}{(x+3)(x-2)} = \dfrac{x^2 + 5x - 19}{(x+3)(x-2)}$

Add the following rational expressions.

1. $\dfrac{x^2 - 9x - 22}{x^2 - 8x - 33} + \dfrac{x^2 + 12x + 35}{x^2 + 10x + 21}$

2. $\dfrac{1}{x^3} + \dfrac{x-9}{x^2}$

3. $\dfrac{1}{2x} + \dfrac{x^2 - 3}{2x}$

4. $\dfrac{y^3}{x} + \dfrac{x^3}{y}$

5. $\dfrac{(x-2)(x+13)}{x^2 - 4} + \dfrac{(x-3)(x+1)}{x^2 - x - 6}$

6. $\dfrac{a}{b^2} + \dfrac{b^3}{a-b}$

7. $\dfrac{5}{y^3} + \dfrac{7}{y^3}$

8. $\dfrac{12b}{a^2} + \dfrac{31c}{a^2}$

9. $\dfrac{y}{x^2} + \dfrac{3}{y^2} + \dfrac{x}{x^2 y^2}$

10. $\dfrac{r}{q^{12}} + \dfrac{s}{r^{15}} + \dfrac{q}{s^5}$

11. $\dfrac{x}{x-y} + \dfrac{y}{x-y}$

12. $\dfrac{9b^{-1}}{a} + \dfrac{2}{b}$

7.9 Subtracting Rational Expressions

Subtracting rational expressions is similar to subtracting fractions because a rational expression is a fraction.

Example 14: Subtract: $\dfrac{x^2}{5} - \dfrac{9}{x+3}$

Step 1: Just like adding we must find a common denominator. The common denominator of the rational expressions is $5(x+3)$.

Step 2: Set up the algebra problem like a fraction problem and find the numerators.

$$\dfrac{x^2}{5} = \dfrac{x^2(x+3)}{5(x+3)}$$
$$-\dfrac{9}{x+3} = \dfrac{45}{5(x+3)}$$

Step 3: Subtract: $\dfrac{x^2(x+3) - 45}{5(x+3)}$

Step 4: Simplify: $\dfrac{x^2(x+3) - 45}{5(x+3)} = \dfrac{x^3 + 3x^2 - 45}{5(x+3)}$

Subtract the following rational expressions.

1. $\dfrac{x}{3y} - \dfrac{2}{3y}$

2. $\dfrac{7}{5c} - \dfrac{2}{c}$

3. $\dfrac{x}{y^2} - \dfrac{x^2}{y^2}$

4. $\dfrac{x+1}{y+1} - \dfrac{y+1}{x+1}$

5. $\dfrac{1}{x} - \dfrac{2x^2}{y}$

6. $\dfrac{2a^2}{bc} - \dfrac{a^5}{bc}$

7. $\dfrac{d-e}{a+b} - \dfrac{5}{a+b}$

8. $\dfrac{x}{5y^2} - \dfrac{y}{5x^2}$

9. $\dfrac{c^3}{3b} - \dfrac{b^2}{4a}$

10. $\dfrac{2x+5}{y} - \dfrac{5x^3 - x^2}{y}$

11. $\dfrac{2d}{3c^5} - \dfrac{a^5}{2b^2}$

12. $\dfrac{5x}{y} - \dfrac{2x}{y}$

Chapter 7   Factoring

7.10   Multiplying Rational Expressions

Multiplying rational expressions is similar to multiplying fractions because a rational expression is a fraction.

Example 15:   Multiply: $\dfrac{x^4 - x^3}{y^3} \times \dfrac{y^4}{x^2}$

Step 1:   When you multiply a rational expression, you must multiply the numerators together and multiply the denominators together.
$$\dfrac{x^4 - x^3}{y^3} \times \dfrac{y^4}{x^2} = \dfrac{(x^4 - x^3) \times y^4}{y^3 \times x^2} = \dfrac{x^4 y^4 - x^3 y^4}{x^2 y^3}$$

Step 2:   Simplify the resulting rational expression. You can factor $x^3 y^4$ out of the numerator.
$$\dfrac{x^4 y^4 - x^3 y^4}{x^2 y^3} = \dfrac{x^3 y^4 (x - 1)}{x^2 y^3}$$

Step 3:   You can also cancel $x^2 y^3$ because it is in the numerator and denominator of the expression.
$$\dfrac{x^3 y^4 (x - 1)}{x^2 y^3} = \dfrac{xy (x - 1)}{1} = xy (x - 1)$$
Therefore, $\dfrac{x^4 - x^3}{y^3} \times \dfrac{y^4}{x^2} = xy (x - 1)$.

Multiply the following rational expressions.

1. $\dfrac{-b}{2a} \times \dfrac{-a}{3b}$

2. $\dfrac{4x}{y} \times \dfrac{1}{2y}$

3. $\dfrac{5a}{3b} \times \dfrac{4c}{3b}$

4. $\dfrac{a+b}{a-b} \times \dfrac{c+b}{c-b}$

5. $\dfrac{c}{a^2} \times \dfrac{c-a}{b-c}$

6. $\dfrac{y^5}{x^3} \times \dfrac{y^2 + 2y + 1}{x^2 - y}$

7. $\dfrac{x^2 - 2x - 3}{x^2 - 5x - 14} \times \dfrac{x^2 - 2x - 35}{x^2 + 6x - 27}$

8. $\dfrac{9}{x} \times \dfrac{x^5}{y}$

9. $\dfrac{a^2 - b}{a - b^2} \times \dfrac{7c}{b}$

10. $\dfrac{5}{x^2 y} \times \dfrac{4}{xy^2}$

11. $\dfrac{5x^3}{2} \times \dfrac{2x^3}{5}$

12. $\dfrac{x+7}{x-1} \times \dfrac{x-3}{x+5}$

13. $\dfrac{3c}{a^2} \times \dfrac{2ba}{c}$

14. $\dfrac{a^5 - a^2}{b^3} \times \dfrac{c^4}{a^2}$

15. $\dfrac{b-17}{c^2 + 2} \times \dfrac{a^3}{b^3} \times \dfrac{c^3}{b-17}$

16. $\dfrac{hk}{47} \times \dfrac{16}{m}$

7.11 Dividing Rational Expressions

Dividing rational expressions is similar to dividing fractions because a rational expression is a fraction.

Example 16: Divide: $\dfrac{2a^3}{b} \div \dfrac{5}{b^2}$

Step 1: Just like dividing fractions, you must flip the second expression to get it's reciprocal, then multiply.
$$\dfrac{2a^3}{b} \div \dfrac{5}{b^2} = \dfrac{2a^3}{b} \times \dfrac{b^2}{5}$$

Step 2: Multiply the two rational expressions together just like you did in the previous section.
$$\dfrac{2a^3}{b} \times \dfrac{b^2}{5} = \dfrac{2a^3 \times b^2}{b \times 5} = \dfrac{2a^3 b^2}{5b}$$

Step 3: Simplify: $\dfrac{2a^3 b^2}{5b} = \dfrac{2a^3 b}{5}$

Therefore, $\dfrac{2a^3}{b} \div \dfrac{5}{b^2} = \dfrac{2a^3 b}{5}$.

Divide the following rational expressions.

1. $\dfrac{x}{y} \div \dfrac{y}{x}$

2. $\dfrac{2x}{y^3} \div \dfrac{x}{y}$

3. $\dfrac{x-1}{y+2} \div \dfrac{y-5}{x+7}$

4. $\dfrac{a-2}{c-1} \div \dfrac{a-2}{c+3}$

5. $\dfrac{x^2+2x+1}{y^2+8y+15} \div \dfrac{x^2+10x+9}{y^2+10y+21}$

6. $\dfrac{a}{c} \div \dfrac{c}{b}$

7. $\dfrac{c}{a} \div \dfrac{c}{b}$

8. $\dfrac{2x+3}{x+3} \div \dfrac{x+3}{2x+3} \div \dfrac{x}{y}$

9. $\dfrac{ac}{b^2} \div \dfrac{c^2}{ab}$

10. $\dfrac{b^2}{a} \div \dfrac{a^2}{c}$

11. $\dfrac{c^2}{3b} \div \dfrac{16}{a}$

12. $\dfrac{20cd}{b} \div \dfrac{2ac}{d}$

13. $\dfrac{x+4}{x+2} \div \dfrac{x-1}{x-3}$

14. $\dfrac{y}{3x^2} \div \dfrac{2y}{x^3}$

15. $\dfrac{y^3-1}{x^2+1} \div \dfrac{x-7}{y+2}$

16. $\dfrac{200a}{7b} \div \dfrac{10a}{b}$

Chapter 7   Factoring

Chapter 7 Review

Factor the following polynomials completely.

1. $8x - 18$
2. $6x^2 - 18x$
3. $16b^3 + 8b$
4. $15a^3 + 40$
5. $20y^6 - 12y^4$
6. $5a - 15a^2$
7. $4y^2 - 36$
8. $25a^4 - 49b^2$
9. $3ax + 3ay + 4x + 4y$
10. $ax - 2x + ay - 2y$
11. $2bx + 2x - 2by - 2y$
12. $2b^2 - 2b - 12$
13. $yx^3 + 14x - 3x^2 - 6$
14. $3a^3 + 4a^2 + 9a + 12$
15. $27y^2 + 42y - 5$
16. $12b^2 + 25b - 7$
17. $c^2 + cd - 20d^2$
18. $x^2 - 4xy - 21y^2$
19. $6y^2 + 30y + 36$
20. $2b^2 + 6b - 20$
21. $16b^4 - 81d^4$
22. $9w^2 - 54w - 63$
23. $m^2p^2 - 5mp + 2m^2p - 10m$
24. $12x^2 + 27x$
25. $2xy - 36 + 8y - 9x$
26. $2a^4 - 32$
27. $21c^2 + 41c + 10$
28. $x^2 - y + xy - x$
29. $2b^3 - 24 + 16b - 3b^2$
30. $5 - 2a - 25a^2 + 10a^3$

Simplify the following rational expressions by performing the appropriate operation.

31. $\dfrac{a^{-1}}{b^{-1}} + \dfrac{b}{a}$

32. $\dfrac{x}{y} + \dfrac{y}{x}$

33. $\dfrac{x}{x+y} - \dfrac{y}{x+y}$

34. $\dfrac{3x+5}{x+1} - \dfrac{3x-1}{x-1}$

35. $\dfrac{-b}{ac} \times \dfrac{3b}{a-2} \times \dfrac{c}{b+1}$

36. $\dfrac{x}{y^3} \times \dfrac{3y^2}{x^2}$

37. $\dfrac{-b}{5ac} \div \dfrac{d}{b^2}$

38. $\dfrac{5}{a+3} \div \dfrac{7a+21}{b^2+b}$

Chapter 7 Test

1. What is the greatest common factor of $4x^3$ and $8x^2$?

   (A) $4x^2$
   (B) $4x$
   (C) x^2
   (D) $8x$

2. Factor: $8x^4 - 7x^2 + 4x$

   (A) $4x(2x^3 - 7x + 4)$
   (B) $x(8x^4 - 7x^2 + 4x)$
   (C) $x(8x^3 - 7x + 4)$
   (D) $4x(2x^3 - 7x + 1)$

3. Factor by grouping:

   $xy + 2x + 3y + 6$

   (A) $(x+2)(y+3)$
   (B) $(x+3)(y+2)$
   (C) $x(y+2) + 3(y+2)$
   (D) This problem cannot be factored.

4. Factor: $x^2 + 6x + 8$

   (A) $(x+2)(x+4)$
   (B) $(x+1)(x+8)$
   (C) $(x-2)(x-4)$
   (D) $(x-1)(x-8)$

5. Divide: $\dfrac{5}{2x} \div \dfrac{5x}{y}$

   (A) $\dfrac{y}{2x}$
   (B) $\dfrac{25}{2x^2 y}$
   (C) $\dfrac{y}{2x^2}$
   (D) $\dfrac{25x}{2y}$

6. Simplify the following rational expression:

   $\dfrac{36x^4 - 16}{6x^3 + 4x}$

   (A) $\dfrac{(6x^2 - 4)(6x^2 + 4)}{x}$
   (B) $\dfrac{6x^2 - 4}{x}$
   (C) $6x^2 - 4$
   (D) $\dfrac{(3x-2)(3x+2)}{x}$

7. Simplify the following rational expression:

   $\dfrac{c^2 + 10c + 24}{c^3 + 4c^2}$

   (A) $\dfrac{c+6}{c}$
   (B) $\dfrac{c+4}{c^2}$
   (C) $\dfrac{c+6}{c^2}$
   (D) $\dfrac{(c+6)(c+4)}{c^2(c+4)}$

8. Add: $\dfrac{3x^3}{y} + \dfrac{3y^3}{x}$

   (A) $\dfrac{9x^3 y^3}{xy}$
   (B) $\dfrac{3x^4 + 3y^4}{xy}$
   (C) $\dfrac{3x^3 + 3y^3}{x+y}$
   (D) $\dfrac{3x^4 + 3y^4}{x+y}$

Chapter 7 Factoring

9. Add: $\dfrac{11x-y}{y} + \dfrac{2}{xy^2}$

(A) $\dfrac{11x^2y - xy^2 + 2}{xy^3}$

(B) $\dfrac{11x^2y - xy^2 - 2}{xy^2}$

(C) $\dfrac{22x - 2y}{y + xy^2}$

(D) $\dfrac{11x^2y - xy^2 + 2}{xy^2}$

10. Subtract: $\dfrac{x+2}{x^2-4} - \dfrac{x}{x-2}$

(A) $\dfrac{1-x}{x-2}$

(B) $\dfrac{x}{x^2-4}$

(C) $\dfrac{2x-2}{x^2-4}$

(D) $\dfrac{x+2}{x^2-x-6}$

11. Multiply: $\dfrac{x+3}{2x} \times \dfrac{x(y-1)}{y}$

(A) $\dfrac{xy - x + 3y - 3}{2x}$

(B) $\dfrac{xy - x + 3y - 3}{2xy}$

(C) $\dfrac{xy - x + 3y - 3}{xy}$

(D) $\dfrac{xy - x + 3y - 3}{2y}$

12. Factor: $2x^2 - 2x - 84$

(A) $(2x+7)(x-12)$
(B) $(2x-12)(x+7)$
(C) $(2x-7)(x+12)$
(D) $(2x+12)(x-7)$

13. Subtract: $\dfrac{1}{7x} - \dfrac{13y}{x}$

(A) $\dfrac{14y}{8x}$

(B) $-\dfrac{13y}{7x}$

(C) $\dfrac{1-13y}{6x}$

(D) $\dfrac{1-91y}{7x}$

14. Multiply: $\dfrac{12y}{x} \times \dfrac{y^3}{x}$

(A) $\dfrac{12y^4}{x^2}$

(B) $\dfrac{12y}{x^3}$

(C) $\dfrac{36y^4}{x^2}$

(D) $\dfrac{12y^4}{x}$

15. Divide: $\dfrac{x-2}{y} \div \dfrac{7}{x+3}$

(A) $\dfrac{7x-14}{y+x+3}$

(B) $\dfrac{x^2+x-6}{7y}$

(C) $\dfrac{7x-14}{xy+3y}$

(D) $\dfrac{x^2+x-6}{y}$

16. Factor: $4x^2 - 64$

(A) $(x-8)(x+8)$
(B) $(4x-8)(4x+8)$
(C) $(2x-16)(2x+16)$
(D) $(2x-8)(2x+8)$

Chapter 8
Solving Quadratic Equations

This chapter covers the following CA Integrated Math I standards:

Algebra	14.0
	19.0
	20.0

In the previous chapter, we factored polynomials such as $y^2 - 4y - 5$ into two factors:

$$y^2 - 4y - 5 = (y+1)(y-5)$$

In this chapter, we learn that any equation that can be put in the form $ax^2 + bx + c = 0$ is a quadratic equation if a, b, and c are real numbers and $a \neq 0$. $ax^2 + bx + c = 0$ is the standard form of a quadratic equation. To solve these equations, follow the steps below.

Example 1: Solve $y^2 - 4y - 5 = 0$

Step 1: Factor the left side of the equation.

$$y^2 - 4y - 5 = 0$$
$$(y+1)(y-5) = 0$$

Step 2: If the product of these two factors equals zero, then the two factors individually must be equal to zero. Therefore, to solve, we set each factor equal to zero.

$$\begin{array}{rl} (y+1) &= 0 \\ -1 & -1 \\ \hline y &= -1 \end{array} \qquad \begin{array}{rl} (y-5) &= 0 \\ +5 & +5 \\ \hline y &= 5 \end{array}$$

The equation has two solutions: $y = -1$ and $y = 5$

Check: To check, substitute each solution into the original equation.

When $y = -1$, the equation becomes:
$(-1)^2 - (4)(-1) - 5 = 0$
$1 + 4 - 5 = 0$
$0 = 0$

When $y = 5$, the equation becomes:
$5^2 - (4)(5) - 5 = 0$
$25 - 20 - 5 = 0$
$0 = 0$

Both solutions produce true statements.
The solution set for the equation is $\{-1, -5\}$

Copyright ©American Book Company

Chapter 8   Solving Quadratic Equations

Solve each of the following quadratic equations by factoring and setting each factor equal to zero. Check by substituting answers back in the original equation.

1. $x^2 + x - 6 = 0$
2. $y^2 - 2y - 8 = 0$
3. $a^2 + 2a - 15 = 0$
4. $y^2 - 5y + 4 = 0$
5. $b^2 - 9b + 14 = 0$
6. $x^2 - 3x - 4 = 0$
7. $y^2 + y - 20 = 0$
8. $d^2 + 6d + 8 = 0$
9. $y^2 - 7y + 12 = 0$
10. $x^2 - 3x - 28 = 0$
11. $a^2 - 5a + 6 = 0$
12. $b^2 + 3b - 10 = 0$
13. $a^2 + 7a - 8 = 0$
14. $c^2 + 3x + 2 = 0$
15. $x^2 - x - 42 = 0$
16. $a^2 + a - 6 = 0$
17. $b^2 + 7b + 12 = 0$
18. $y^2 + 2y - 15 = 0$
19. $a^2 - 3a - 10 = 0$
20. $d^2 + 10d + 16 = 0$
21. $x^2 - 4x - 12 = 0$

Quadratic equations that have a whole number and a variable in the first term are solved the same way as the previous page. Factor the trinomial, and set each factor equal to zero to find the solution set.

Example 2:   Solve $2x^2 + 3x - 2 = 0$
$(2x - 1)(x + 2) = 0$
Set each factor equal to zero and solve:

$$2x - 1 = 0$$
$$+1 \quad +1$$
$$\frac{2x}{2} = \frac{1}{2}$$
$$x = \frac{1}{2}$$

$$x + 2 = 0$$
$$-2 \quad -2$$
$$x = -2$$

The solution set is $\left\{\frac{1}{2}, -2\right\}$

Solve the following quadratic equations.

22. $3y^2 + 4y - 32 = 0$
23. $5c^2 - 2c - 16 = 0$
24. $7d^2 + 18d + 8 = 0$
25. $3a^2 - 10a - 8 = 0$
26. $11x^2 - 31x - 6 = 0$
27. $5b^2 + 17b + 6 = 0$
28. $3x^2 - 11x - 20 = 0$
29. $5a^2 + 47a - 30 = 0$
30. $2c^2 - 5c - 25 = 0$
31. $2y^2 + 11y - 21 = 0$
32. $5a^2 + 23a - 42 = 0$
33. $3d^2 + 11d - 20 = 0$
34. $3x^2 - 10x + 8 = 0$
35. $7b^2 + 23b - 20 = 0$
36. $9a^2 - 58a + 24 = 0$
37. $4c^2 - 25c - 21 = 0$
38. $8d^2 + 53d + 30 = 0$
39. $4y^2 - 29x + 30 = 0$
40. $8a^2 + 37a - 15 = 0$
41. $3x^2 - 41x + 26 = 0$
42. $8b^2 + 2b - 3 = 0$

8.1 Solving the Difference of Two Squares

To solve the difference of two squares, first factor. Then set each factor equal to zero.

Example 3: $25x^2 - 36 = 0$

Step 1: Factor the left hand side of the equation.

$25x^2 - 36 = 0$
$(5x + 6)(5x - 6) = 0$

Step 2: Set each factor equal to zero and solve.

$$\begin{aligned} 5x + 6 &= 0 \\ -6 & -6 \\ \hline \frac{5x}{5} &= \frac{6}{5} \\ x &= -\frac{6}{5} \end{aligned} \qquad \begin{aligned} 5x - 6 &= 0 \\ +6 & +6 \\ \hline \frac{5x}{5} &= \frac{6}{5} \\ x &= \frac{6}{5} \end{aligned}$$

Check: Substitute each solution in the equation to check.

for $x = -\frac{6}{5}$:

$25x^2 - 36 = 0$

$25\left(-\frac{6}{5}\right)\left(-\frac{6}{5}\right) - 36 = 0 \longleftarrow$ Substitute $-\frac{6}{5}$ for x.

$25\left(\frac{36}{25}\right) - 36 = 0 \longleftarrow$ Cancel the 25's.

$36 - 36 = 0 \longleftarrow$ A true statement. $x = -\frac{6}{5}$ is a solution.

for $x = \frac{6}{5}$:

$25x^2 - 36 = 0$

$25\left(\frac{6}{5}\right)\left(\frac{6}{5}\right) - 36 = 0 \longleftarrow$ Substitute $\frac{6}{5}$ for x.

$25\left(\frac{36}{25}\right) - 36 = 0 \longleftarrow$ Cancel the 25's.

$36 - 36 = 0 \longleftarrow$ A true statement. $x = \frac{6}{5}$ is a solution.

The solution set is $\left\{\frac{-6}{5}, \frac{6}{5}\right\}$.

Chapter 8 Solving Quadratic Equations

Find the solution sets for the following.

1. $25a^2 - 16 = 0$

2. $c^2 - 36 = 0$

3. $9x^2 - 64 = 0$

4. $100y^2 - 49 - 0$

5. $4b^2 - 81 = 0$

6. $d^2 - 25 = 0$

7. $9x^2 - 1 = 0$

8. $16a^2 - 9 = 0$

9. $36y^2 - 1 = 0$

10. $36y^2 - 25 = 0$

11. $d^2 - 16 = 0$

12. $64b^2 - 9 = 0$

13. $81a^2 - 4 = 0$

14. $64y^2 - 25 = 0$

15. $4c^2 - 49 = 0$

16. $x^2 - 81 = 0$

17. $49b^2 - 9 = 0$

18. $a^2 - 64 = 0$

19. $x^2 - 1 = 0$

20. $4y^2 - 9 = 0$

21. $t^2 - 100 = 0$

22. $16k^2 - 81 = 0$

23. $a^2 - 4 = 0$

24. $36b^2 - 16 = 0$

8.2 Solving Perfect Squares

When the square root of a constant, variable, or polynomial results in a constant, variable, or polynomial without irrational numbers, the expression is a **perfect square**. Some examples are 49, x^2, and $(x-2)^2$.

Example 4: Solve the perfect square for x. $(x-5)^2 = 0$

Step 1: Take the square root of both sides.
$\sqrt{(x-5)^2} = \sqrt{0}$
$(x-5) = 0$

Step 2: Solve the equation.
$(x-5) = 0$
$x - 5 + 5 = 0 + 5$
$x = 5$

Example 5: Solve the perfect square for x. $(x-5)^2 = 64$

Step 1: Take the square root of both sides.
$\sqrt{(x-5)^2} = \sqrt{64}$
$(x-5) = \pm 8$
$(x-5) = 8$ and $(x-5) = -8$

Step 2: Solve the two equations.
$(x-5) = 8$ and $(x-5) = -8$
$x - 5 + 5 = 8 + 5$ and $x - 5 + 5 = -8 + 5$
$x = 13$ and $x = -3$

Solve the perfect square for x.

1. $(x-2)^2 = 0$
2. $(x+1)^2 = 0$
3. $(x+11)^2 = 0$
4. $(x-4)^2 = 0$
5. $(x-1)^2 = 0$
6. $(x+8)^2 = 0$
7. $(x+3)^2 = 4$
8. $(x-5)^2 = 16$
9. $(x-10)^2 = 100$
10. $(x+9)^2 = 9$
11. $(x-4.5)^2 = 25$
12. $(x+7)^2 = 36$
13. $(x+2)^2 = 49$
14. $(x-1)^2 = 4$
15. $(x+8.9)^2 = 49$
16. $(x-6)^2 = 81$
17. $(x-12)^2 = 121$
18. $(x+2.5)^2 = 64$

8.3 Completing the Square

"Completing the Square" is another way of factoring a quadratic equation. To complete the square, convert the equation into a perfect square.

Example 6: Solve $x^2 - 10x + 9 = 0$ by completing the square.

Completing the square:

Step 1: The first step is to get the constant on the other side of the equation. Subtract 9 from both sides:
$x^2 - 10x + 9 - 9 = -9$
$x^2 - 10x = -9$

Step 2: Determine the coefficient of the x. The coefficient in this example is 10. Divide the coefficient by 2 and square the result.
$(10 \div 2)^2 = 5^2 = 25$

Step 3: Add the resulting value, 25, to both sides:
$x^2 - 10x + 25 = -9 + 25$
$x^2 - 10x + 25 = 16$

Step 4: Now factor the $x^2 - 10x + 25$ into a perfect square:
$(x - 5)^2 = 16$

Solving the perfect square:

Step 5: Take the square root of both sides.
$\sqrt{(x-5)^2} = \sqrt{16}$
$(x - 5) = \pm 4$
$(x - 5) = 4$ and $(x - 5) = -4$

Step 6: Solve the two equations.
$(x - 5) = 4$ and $(x - 5) = -4$
$x - 5 + 5 = 4 + 5$ and $x - 5 + 5 = -4 + 5$
$x = 9$ and $x = 1$

Solve for x by completing the square.

1. $x^2 + 2x - 3 = 0$
2. $x^2 - 8x + 7 = 0$
3. $x^2 + 6x - 7 = 0$
4. $x^2 - 16x - 36 = 0$
5. $x^2 - 14x + 49 = 0$
6. $x^2 - 4x = 0$
7. $x^2 + 12x + 27 = 0$
8. $x^2 + 2x - 24 = 0$
9. $x^2 + 12x - 85 = 0$
10. $x^2 - 8x + 15 = 0$
11. $x^2 - 16x + 60 = 0$
12. $x^2 - 8x - 48 = 0$
13. $x^2 + 24x + 44 = 0$
14. $x^2 + 6x + 5 = 0$
15. $x^2 - 11x + 5.25 = 0$

8.4 Proof of the Quadratic Formula

The quadratic formula $\dfrac{-b \pm \sqrt{b^2 - 4ac}}{2a}$ can be proved by using the "completing the square" method on the quadratic equation $ax^2 + bx + c = 0$.

$$ax^2 + bx + c = 0$$

$$ax^2 + bx + c - c = 0 - c \qquad \text{Subtract } c \text{ from both sides.}$$

$$ax^2 + bx = -c \qquad \text{Simplify.}$$

$$\frac{ax^2 + bx}{a} = \frac{-c}{a} \qquad \text{Divide } a \text{ on both sides.}$$

$$x^2 + \frac{bx}{a} = \frac{-c}{a} \qquad \text{Simplify.}$$

$$x^2 + \frac{bx}{a} + \left(\frac{b}{2a}\right)^2 = \frac{-c}{a} + \left(\frac{b}{2a}\right)^2 \qquad \text{Complete the square by adding } \left(\frac{b}{2a}\right)^2 \text{ to both sides.}$$

$$\left(x + \frac{b}{2a}\right)^2 = \frac{-c}{a} + \left(\frac{b}{2a}\right)^2 \qquad \text{Factor the left side.}$$

$$\left(x + \frac{b}{2a}\right)^2 = \frac{-c}{a} + \frac{b^2}{4a^2} \qquad \text{Square } \frac{b}{2a} \text{ on the right side.}$$

$$\left(x + \frac{b}{2a}\right)^2 = \frac{-4ac}{4a^2} + \frac{b^2}{4a^2} \qquad \text{Find a common denominator on the right side so the fractions can be added.}$$

$$\left(x + \frac{b}{2a}\right)^2 = \frac{b^2 - 4ac}{4a^2} \qquad \text{Add the fractions on the right side.}$$

$$\sqrt{\left(x + \frac{b}{2a}\right)^2} = \sqrt{\frac{b^2 - 4ac}{4a^2}} \qquad \text{Take the square root of both sides.}$$

$$x + \frac{b}{2a} = \frac{\pm\sqrt{b^2 - 4ac}}{2a} \qquad \text{Simplify.}$$

$$x + \frac{b}{2a} - \frac{b}{2a} = \frac{\pm\sqrt{b^2 - 4ac}}{2a} - \frac{b}{2a} \qquad \text{Subtract } \frac{b}{2a} \text{ from both sides.}$$

$$x = \frac{-b \pm \sqrt{b^2 - 4ac}}{2a} \qquad \text{Add the fractions. The proof is complete.}$$

Copyright © American Book Company

Chapter 8 Solving Quadratic Equations

8.5 Using the Quadratic Formula

You may be asked to use the quadratic formula to solve an algebra problem known as a **quadratic equation**. The equation should be in the form $ax^2 + bx + c = 0$.

Example 7: Using the quadratic formula, find x in the following equation: $x^2 - 8x = -7$.

Step 1: Make sure the equation is set equal to 0.

$$x^2 - 8x + 7 = -7 + 7$$
$$x^2 - 8x + 7 = 0$$

The quadratic formula, $\dfrac{-b \pm \sqrt{b^2 - 4ac}}{2a}$, will be given to you on your formula sheet with your test.

Step 2: In the formula, a is the number x^2 is multiplied by, b is the number x is multiplied by and c is the last term of the equation. For the equation in the example, $x^2 - 8x + 7$, $a = 1$, $b = -8$, and $c = 7$. When we look at the formula we notice a $\pm$ sign. This means that there will be two solutions to the equation, one when we use the plus sign and one when we use the minus sign. Substituting the numbers from the problem into the formula, we have:

$$\dfrac{8 + \sqrt{8^2 - (4)(1)(7)}}{2(1)} = 7 \quad \text{or} \quad \dfrac{8 - \sqrt{8^2 - (4)(1)(7)}}{2(1)} = 1$$

The solutions are $\{7, 1\}$

For each of the following equations, use the quadratic formula to find two solutions.

1. $x^2 + x - 6 = 0$
2. $y^2 - 2y - 8 = 0$
3. $a^2 + 2a - 15 = 0$
4. $y^2 - 5y + 4 = 0$
5. $b^2 - 9b + 14 = 0$
6. $x^2 - 3x - 4 = 0$
7. $y^2 + y - 20 = 0$
8. $d^2 + 6d + 8 = 0$
9. $y^2 - 7y + 12 = 0$
10. $x^2 - 3x - 28 = 0$
11. $a^2 - 5a + 6 = 0$
12. $b^2 + 3b - 10 = 0$
13. $a^2 + 7a - 8 = 0$
14. $c^2 + 3c + 2 = 0$
15. $x^2 - x - 42 = 0$
16. $a^2 + 5a - 6 = 0$
17. $b^2 + 7b + 12 = 0$
18. $y^2 + y - 12 = 0$
19. $a^2 - 3a - 10 = 0$
20. $d^2 + 10d + 16 = 0$
21. $x^2 - 4x - 12 = 0$

Chapter 8 Review

Factor and solve each of the following quadratic equations.

1. $16b^2 - 25 = 0$

2. $a^2 - a - 30 = 0$

3. $x^2 - x = 6$

4. $100x^2 - 49 = 0$

5. $81y^2 = 9$

6. $y^2 = 21 - 4y$

7. $y^2 - 7y + 8 = 16$

8. $6x^2 + x - 2 = 0$

9. $3y^2 + y - 2 = 0$

10. $b^2 + 2b - 8 = 0$

11. $4x^2 + 19x - 5 = 0$

12. $8x^2 = 6x + 2$

13. $2y^2 - 6y - 20 = 0$

14. $-6x^2 + 7x - 2 = 0$

15. $y^2 + 3y - 18 = 0$

Using the quadratic formula, find both solutions for the variable.

16. $x^2 + 10x - 11 = 0$

17. $y^2 - 14y + 40 = 0$

18. $b^2 + 9b + 18 = 0$

19. $y^2 - 12y - 13 = 0$

20. $a^2 - 8a - 48 = 0$

21. $x^2 + 2x - 63 = 0$

Chapter 8 Solving Quadratic Equations

Chapter 8 Test

1. Solve: $4y^2 - 9y = -5$

   (A) $\{1, \frac{5}{4}\}$

   (B) $\{-\frac{3}{4}, -1\}$

   (C) $\{-1, \frac{4}{5}\}$

   (D) $\{\frac{5}{16}, 1\}$

2. Solve for y: $2y^2 + 13y + 15 = 0$

   (A) $\{\frac{3}{2}, \frac{5}{2}\}$

   (B) $\{\frac{2}{3}, \frac{2}{5}\}$

   (C) $\{-5, -\frac{3}{2}\}$

   (D) $\{5, -\frac{3}{2}\}$

3. Solve for x.

   $x^2 - 3x - 18$

   (A) $\{-6, 3\}$
   (B) $\{6, -3\}$
   (C) $\{-9, 2\}$
   (D) $\{9, -2\}$

4. What are the values of x in the quadratic equation?

   $x^2 + 2x - 15 = x - 3$

   (A) $\{-4, 3\}$
   (B) $\{-3, 4\}$
   (C) $\{-3, 5\}$
   (D) Cannot be determined

5. Solve the equation $(x+9)^2 = 49$

   (A) $x = -9, 9$
   (B) $x = -9, 7$
   (C) $x = -16, -2$
   (D) $x = -7, 7$

6. Solve the equation $c^2 + 8c - 9 = 0$ by completing the square.

   (A) $c = \{1, -9\}$
   (B) $c = \{-1, 9\}$
   (C) $c = \{3, 3\}$
   (D) $c = \{-3, -3\}$

7. Using the quadratic formula, solve the following equation:

   $3x^2 = 9x$

   (A) $x = \{0, 1\}$
   (B) $x = \{3, 1\}$
   (C) $x = \{0, 3\}$
   (D) $x = \{3, -3\}$

8. Solve $6a^2 + 11a - 10 = 0$, using the quadratic formula.

   (A) $\{-\frac{2}{5}, \frac{3}{2}\}$

   (B) $\{\frac{2}{5}, \frac{2}{3}\}$

   (C) $\{-\frac{5}{2}, \frac{2}{3}\}$

   (D) $\{\frac{5}{2}, \frac{2}{3}\}$

Chapter 9
Graphing and Writing Equations and Inequalities

This chapter covers the following CA Integrated Math I standards:

Algebra	6.0
	7.0
	8.0
Geometry	17.0

9.1 Forms of Linear Equations

Any equation with two variables that are both to the first power is called a **linear equation.** For example, $x + 3y = 6$. It is called "linear" because the graph of the equation is a straight line on a Cartesian plane. Linear equations come in a number of forms.

The **standard form** of a linear equation is $Ax + By = C$, where $A, B,$ and C are real numbers.

Some examples: $3x + 2y = 6$ $\frac{6}{5}x + \frac{1}{2}y = \frac{2}{7}$ $3.4x - 6.8y = 39.1$ $-x - y = 0$

The **general form** of a linear equation is $Ax + By + D = 0$, where $A, B,$ and D are real numbers.

Some examples: $3x + 2y - 6 = 0$ $\frac{6}{5}x + \frac{1}{2}y - \frac{2}{7} = 0$ $3.4x - 6.8y - 39.1 = 0$ $-x - y = 0$.

Notice that D of standard form equals opposite of C of general form. (It is the same number just opposite signs.)

The **slope-intercept form** of a linear equation is $y = mx + b$, where m is the slope of the line, and b is the y-intercept (the point where the line intersects the y-axis, $x = 0$).

Slope-intercept form is perhaps the most commonly used because m and b are very useful information in many problems.

Some examples: $y = \frac{3}{2}x - 3$ $y = -\frac{12}{5}x + \frac{2}{7}$ $y = 0.5x - 5.75$ $y = -x$

Converting between forms is relatively simple. $4x - y = 8$ is in general form. Subtract 8 from each side: $4x - y - 8 = 8 - 8. \Longrightarrow 4x - y - 8 = 0$ Now the equation is in standard form.
Solve for y by adding y to both sides: $4x - y + y - 8 = 0 + y \Longrightarrow 4x - 8 = y$ is now in slope-intercept form.
You should be able to graph any equation using any of the forms.

Copyright ©American Book Company

Chapter 9   Graphing and Writing Equations and Inequalities

Find the form of the equation of each line.

1. $x + y = 6$
2. $0 = x - 2y + 1$
3. $y = 7x - 2$
4. $x - y + 2 = 0$
5. $9x - 5 = y$
6. $x - 5y = 0$

9.2   Graphing Linear Equations

In addition to graphing ordered pairs, the Cartesian plane can be used to graph the solution set for an equation. Any equation with two variables that are both to the first power is called a **linear equation**. The graph of a linear equation will always be a straight line.

Example 1:   Graph the solution set for $x + y = 7$.

Step 1:   Make a list of some pairs of numbers that will work in the equation.

$$\begin{array}{ll} x + y = 7 & \\ 4 + 3 = 7 & (4, 3) \\ -1 + 8 = 7 & (-1, 8) \\ 5 + 2 = 7 & (5, 2) \\ 0 + 7 = 7 & 0, 7 \end{array} \right\} \text{ordered pair solutions}$$

Step 2:   Plot these points on a Cartesian plane.

Step 3:   By passing a line through these points, we graph the solution set for $x + y = 7$. This means that every point on the line is a solution to the equation $x + y = 7$. For example, $(1, 6)$ is a solution, so the line passes through the point $(1, 6)$.

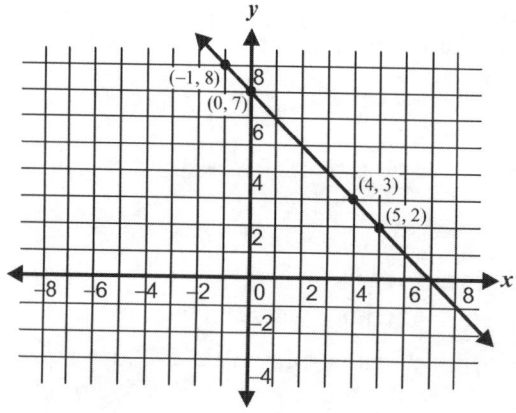

Make a table of solutions for each linear equation below. Then plot the ordered pair solutions on graph paper. Draw a line through the points. (If one of the points does not line up, you have made a mistake.)

1. $x + y = 6$
2. $y = x + 1$
3. $y = x - 2$
4. $x + 2 = y$
5. $x - 5 = y$
6. $x - y = 0$

9.2 Graphing Linear Equations

Example 2: Graph the equation $y = 2x - 5$.

Step 1: This equation has 2 variables, both to the first power, so we know the graph will be a straight line. Substitute some numbers for x or y to find pairs of numbers that satisfy the equation. For the above equation, it will be easier to substitute values of x in order to find the corresponding value for y. Record the values for x and y in a table.

If x is 0, y would be -5
If x is 1, y would be -3
If x is 2, y would be -1
If x is 3, y would be 1

x	y
0	-5
1	-3
2	-1
3	1

Step 2: Graph the ordered pairs, and draw a line through the points.

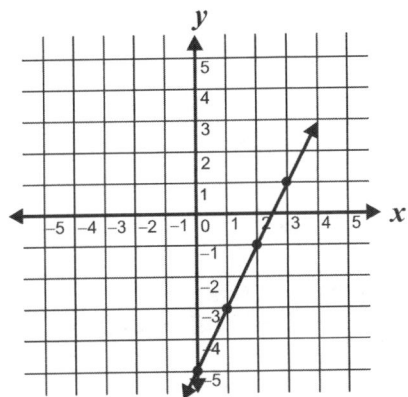

Find pairs of numbers that satisfy the equations below, and graph the line on graph paper.

1. $y = -2x + 2$
2. $2x - 2 = y$
3. $-x + 3 = y$
4. $y = x + 1$
5. $4x - 2 = y$
6. $y = 3x - 3$
7. $x = 4y - 3$
8. $2x = 3y + 1$
9. $x + 2y = 4$

Chapter 9 Graphing and Writing Equations and Inequalities

9.3 Graphing Horizontal and Vertical Lines

The graph of some equations is a horizontal or a vertical line.

Example 3: $y = 3$

Step 1: Make a list of ordered pairs that satisfy the equation $y = 3$.

x	y
0	3
1	3
2	3
3	3

No matter what value of x you choose, y is always 3.

Step 2: Plot these points on an Cartesian plane, and draw a line through the points.

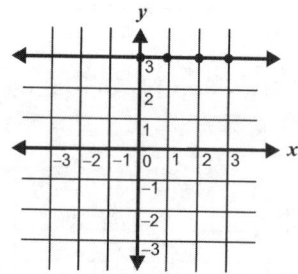

The graph is a horizontal line.

Example 4: $2x + 3 = 0$

Step 1: For these equations with only one variable, find what x equals first.
$2x + 3 = 0$
$2x = -3$
$x = \dfrac{-3}{2}$

Step 2: Using Example 3, find ordered pairs that satisfy the equation, plot the points, and graph the line.

x	y
$\dfrac{-3}{2}$	0
$\dfrac{-3}{2}$	1
$\dfrac{-3}{2}$	2
$\dfrac{-3}{2}$	3

No matter which value of y you choose, the value of x does not change.

The graph is a vertical line.

Find pairs of numbers that satisfy the equations below, and graph the line on graph paper.

1. $2y + 2 = 0$
2. $x = -4$
3. $3x = 3$
4. $y = 5$
5. $4x - 2 = 0$
6. $2x - 6 = 0$
7. $4y = 1$
8. $5x + 10 = 0$
9. $3y + 12 = 0$
10. $x + 1 = 0$
11. $2y - 8 = 0$
12. $3x = -9$
13. $x = -2$
14. $6y - 2 = 0$
15. $5x - 5 = 0$

9.4 Finding the Distance Between Two Points

Notice that a subscript added to the x and y identifies each ordered pair uniquely in the plane. For example, point 1 is identified as (x_1, y_1), point 2 as (x_2, y_2), and so on. This unique subscript identification allows us to calculate slope, distance, and midpoints of line segments in the plane using standard formulas like the distance formula. To find the distance between two points on a Cartesian plane, use the following formula:

$$d = \sqrt{(y_2 - y_1)^2 + (x_2 - x_1)^2}$$

Example 5: Find the distance between $(-2, 1)$ and $(3, -4)$.

Plugging the values from the ordered pairs into the formula, we find:

$$d = \sqrt{(-4 - 1)^2 + [3 - (-2)]^2}$$
$$d = \sqrt{(-5)^2 + (5)^2}$$
$$d = \sqrt{25 + 25} = \sqrt{50}$$

To simplify, we look for perfect squares that are a factor of 50. $50 = 25 \times 2$. Therefore,

$$d = \sqrt{25} \times \sqrt{2} = 5\sqrt{2}$$

Find the distance between the following pairs of points using the distance formula above.

1. $(6, -1) (5, 2)$
2. $(-4, 3) (2, -1)$
3. $(10, 2) (6, -1)$
4. $(-2, 5) (-4, 3)$
5. $(8, -2) (3, -9)$
6. $(2, -2) (8, 1)$
7. $(3, 1) (5, 5)$
8. $(-2, -1) (3, 4)$
9. $(5, -3) (-1, -5)$
10. $(6, 5) (3, -4)$
11. $(-1, 0) (-9, -8)$
12. $(-2, 0) (-6, 6)$
13. $(2, 4) (8, 10)$
14. $(-10, -5) (2, -7)$
15. $(-3, 6) (1, -1)$

9.5 Finding the Midpoint of a Line Segment

You can use the coordinates of the endpoints of a line segment to find the coordinates of the midpoint of the line segment. The formula to find the midpoint between two coordinates is:

$$\text{midpoint}, M = \left(\frac{x_1 + x_2}{2}, \frac{y_1 + y_2}{2}\right)$$

Example 6: Find the midpoint of the line segment having endpoints at $(-3, -1)$ and $(4, 3)$.

Use the formula for the midpoint. $M = \left(\dfrac{4 + (-3)}{2}, \dfrac{3 + (-1)}{2}\right)$

When we simplify each coordinate, we find the midpoint, M, is $\left(\frac{1}{2}, 1\right)$.

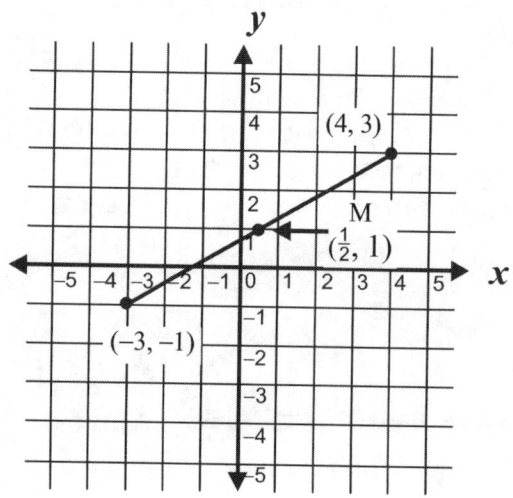

For each of the following pairs of points, find the coordinate of the midpoint, M, using the formula given above.

1. $(4, 5) (-6, 9)$
2. $(-3, 2) (-1, -2)$
3. $(3, 6) (9, 12)$
4. $(2, 5) (6, 9)$
5. $(8, 9) (6, 11)$

6. $(-4, 3) (8, 7)$
7. $(-1, -5) (-3, -11)$
8. $(4, 2) (-2, 8)$
9. $(4, 3) (-1, -5)$
10. $(-6, 2) (8, -8)$

11. $(-3, 9) (-9, 3)$
12. $(7, 8) (11, 6)$
13. $(12, 19) (2, 3)$
14. $(5, 4) (9, -2)$
15. $(-4, 6) (10, -2)$

9.6 Finding the Intercepts of a Line

The x-intercept is the point where the graph of a line crosses the x-axis. The y-intercept is the point where the graph of a line crosses the y-axis.

To find the x-intercept, set $y = 0$
To find the y-intercept, set $x = 0$

Example 7: Find the x- and y-intercepts of the line $6x + 2y = 18$

Step 1: To find the x-intercept, set $y = 0$.

$$\begin{aligned} 6x + 2(0) &= 18 \\ \frac{6x}{6} &= \frac{18}{6} \\ x &= 3 \end{aligned}$$

The x-intercept is at the point $(3, 0)$.

Step 2: To find the y-intercept, set $x = 0$.

$$\begin{aligned} 6(0) + 2y &= 18 \\ \frac{2y}{2} &= \frac{18}{2} \\ y &= 9 \end{aligned}$$

The y-intercept is at the point $(0, 9)$.

You can now use the two intercepts to graph the line.

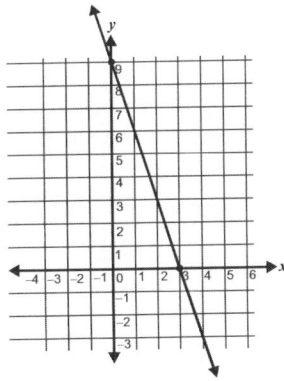

For each of the following equations, find both the x and the y intercepts of the line. For extra practice, draw each of the lines on graph paper.

1. $8x - 2y = 8$
2. $4x + 8y = 16$
3. $3x + 3y = 9$
4. $x - 2y = -5$
5. $8x + 4y = 32$
6. $3x - 4y = 12$
7. $-3x - 3y = 6$
8. $-6x + 2y = 18$
9. $4x - 2y = -4$
10. $-5x - 3y = 15$
11. $3x - 6y = -12$
12. $6x + 3y = 9$
13. $-2x - 6y = 18$
14. $2x + 3y = -6$
15. $-3x + 8y = 12$

Chapter 9 Graphing and Writing Equations and Inequalities

9.7 Understanding Slope

The slope of a line refers to how steep a line is. Slope is also defined as the rate of change. When we graph a line using ordered pairs, we can easily determine the slope. Slope is often represented by the letter m.

$$\text{The formula for slope of a line is: } m = \frac{y_2 - y_1}{x_2 - x_1} \text{ or } \frac{\text{rise}}{\text{run}}$$

Example 8: What is the slope of the following line that passes through the ordered pairs $(-4, -3)$ and $(1, 3)$?

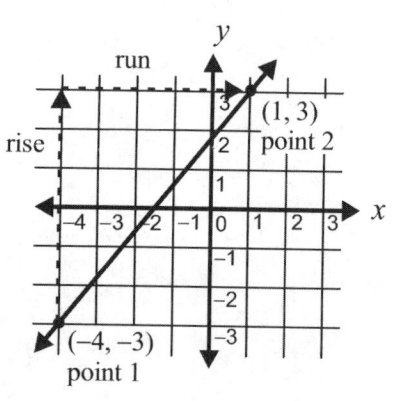

y_2 is 3, the y-coordinate of point 2.

y_1 is -3, the y-coordinate of point 1.

x_2 is 1, the x-coordinate of point 2.

x_1 is -4, the x-coordinate of point 1.

Use the formula for slope given above:

$$m = \frac{3 - (-3)}{1 - (-4)} = \frac{6}{5}$$

The slope is $\frac{6}{5}$. This shows us that we can go up 6 (rise) and over 5 to the right (run) to find another point on the line.

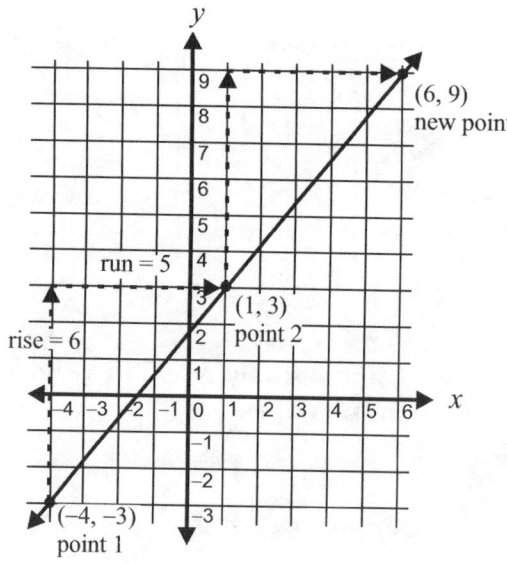

118

9.7 Understanding Slope

Example 9: Find the slope of a line through the points $(-2, 3)$ and $(1, -2)$. It doesn't matter which pair we choose for point 1 and point 2. The answer is the same.

Let point 1 be $(-2, 3)$
Let point 2 be $(1, -2)$

$$\text{slope} = \frac{(y_2 - y_1)}{(x_2 - x_1)} = \frac{-2 - 3}{1 - (-2)} = \frac{-5}{3}$$

When the slope is negative, the line will slant left. For this example, the line will go **down** 5 units and then over 3 units to the **right**.

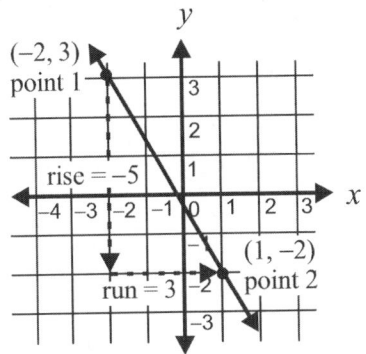

Example 10: What is the slope of a line that passes through $(1, 1)$ and $(3, 1)$?

$$\text{slope} = \frac{1 - 1}{3 - 1} = \frac{0}{2} = 0$$

When $y_2 - y_1 = 0$, the slope will equal 0, and the line will be horizontal.

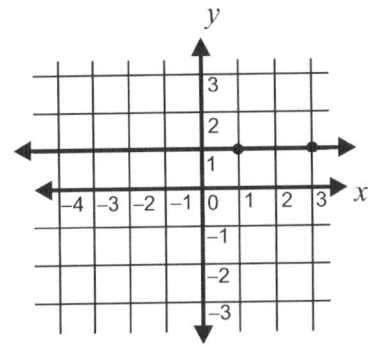

Example 11: What is the slope of a line that passes through $(2, 1)$ and $(2, -3)$?

$$\text{slope} = \frac{-3 - 1}{2 - 2} = \frac{-4}{0} = \text{undefined}$$

When $x_2 - x_1 = 0$, the slope is undefined, and the line will be vertical.

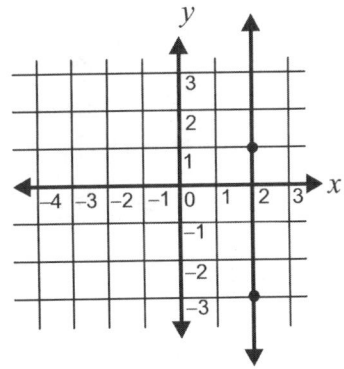

The following lines summarize what we know about slope.

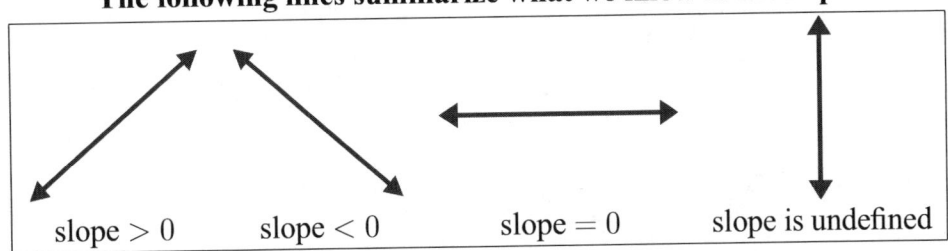

Copyright © American Book Company

Chapter 9   Graphing and Writing Equations and Inequalities

Find the slope of the line that goes through the following pairs of points. Then, using graph paper, graph the line through the two points, and label the rise and run. (See Examples 8–11).

1. $(2, 3)\ (4, 5)$
2. $(1, 3)\ (2, 5)$
3. $(-1, 2)\ (4, 1)$
4. $(1, -2)\ (4, -2)$
5. $(3, 0)\ (3, 4)$
6. $(3, 2)\ (-1, 8)$
7. $(4, 3)\ (2, 4)$
8. $(2, 2)\ (1, 5)$
9. $(3, 4)\ (1, 2)$
10. $(3, 2)\ (3, 6)$
11. $(6, -2)\ (3, -2)$
12. $(1, 2)\ (3, 4)$
13. $(-2, 1)\ (-4, 3)$
14. $(5, 2)\ (4, -1)$
15. $(1, -3)\ (-2, 4)$
16. $(2, -1)\ (3, 5)$

9.8   Slope-Intercept Form of a Line

An equation that contains two variables, each to the first degree, is a **linear equation**. The graph for a linear equation is a straight line. To put a linear equation in slope-intercept form, solve the equation for y. This form of the equation shows the slope and the y-intercept. Slope-intercept form follows the pattern of $y = mx + b$. The "m" represents slope, and the "b" represents the y-intercept. The y-intercept is the point at which the line crosses the y-axis.

When the slope of a line is not 0, the graph of the equation shows a **direct variation** between y and x. When y increases, x increases in a certain proportion. The proportion stays constant. The constant is called the **slope** of the line.

Example 12:   Put the equation $2x + 3y = 15$ in slope-intercept form. What is the slope of the line? What is the y-intercept? Graph the line.

Step 1:   Solve for y:

$$\begin{aligned} 2x + 3y &= 15 \\ -2x & -2x \\ \hline \frac{3y}{3} &= -\frac{2x}{3} + \frac{15}{3} \end{aligned}$$

slope-intercept form: $y = -\frac{2}{3}x + 5$

The slope is $-\frac{2}{3}$ and the y-intercept is 5.

Step 2:   Knowing the slope and the y-intercept, we can graph the line.

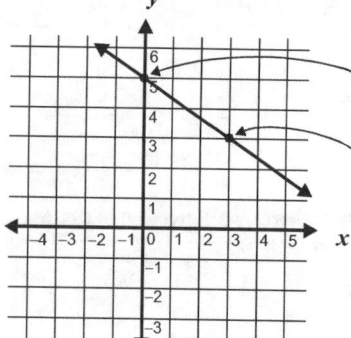

The y-intercept is 5, so the line passes through the point $(0, 5)$ on the y-axis.

The slope is $-\frac{2}{3}$, so go down 2 and over 3 to get a second point.

Put each of the following equations in slope-intercept form by solving for y. On your graph paper, graph the line using the slope and y-intercept.

1. $4x - 5y = 5$
2. $2x + 4y = 16$
3. $3x - 2y = 10$
4. $x + 3y = -12$
5. $6x + 2y = 0$
6. $8x - 5y = 10$
7. $-2x + y = 4$
8. $-4x + 3y = 12$
9. $-6x + 2y = 12$
10. $x - 5y = 5$
11. $3x - 2y = -6$
12. $3x + 4y = 2$
13. $-x = 2 + 4y$
14. $2x = 4y - 2$
15. $6x - 3y = 9$
16. $4x + 2y = 8$
17. $6x - y = 4$
18. $-2x - 4y = 8$
19. $5x + 4y = 16$
20. $6 = 2y - 3x$

9.9 Verify That a Point Lies on a Line

To know whether or not a point lies on a line, substitute the coordinates of the point into the formula for the line. If the point lies on the line, the equation will be true. If the point does not lie on the line, the equation will be false.

Example 13: Does the point $(5, 2)$ lie on the line given by the equation $x + y = 7$?

Solution: Substitute 5 for x and 2 for y in the equation. $5 + 2 = 7$. Since this is a true statement, the point $(5, 2)$ does lie on the line $x + y = 7$.

Example 14: Does the point $(0, 1)$ lie on the line given by the equation $5x + 4y = 16$?

Solution: Substitute 0 for x and 1 for y in the equation $5x + 4y = 16$. Does $5(0) + 4(1) = 16$? No, it equals 4, not 16. Therefore, the point $(0, 1)$ is not on the line given by the equation $5x + 4y = 16$.

For each point below, state whether or not it lies on the line given by the equation that follows the point coordinates.

1. $(2, 4)$ $6x - y = 8$
2. $(1, 1)$ $6x - y = 5$
3. $(3, 8)$ $-2x + y = 2$
4. $(9, 6)$ $-2x + y = 0$
5. $(3, 7)$ $x - 5y = -32$
6. $(0, 5)$ $-6x - 5y = 3$
7. $(2, 4)$ $4x + 2y = 16$
8. $(9, 1)$ $3x - 2y = 29$
9. $(6, 8)$ $6x - y = 28$
10. $(-2, 3)$ $x + 2y = 4$
11. $(4, -1)$ $-x - 3y = -1$
12. $(-1, -3)$ $2x + y = 1$

Chapter 9 Graphing and Writing Equations and Inequalities

9.10 Graphing a Line Knowing a Point and Slope

If you are given a point of a line and the slope of a line, the line can be graphed.

Example 15: Given that line l has a slope of $\frac{4}{3}$ and contains the point $(2, -1)$, graph the line.

Plot and label the point $(2, -1)$ on a Cartesian plane.

The slope, m, is $\frac{4}{3}$, so the rise is 4, and the run is 3. From the point $(2, -1)$, count 4 units up and 3 units to the right.

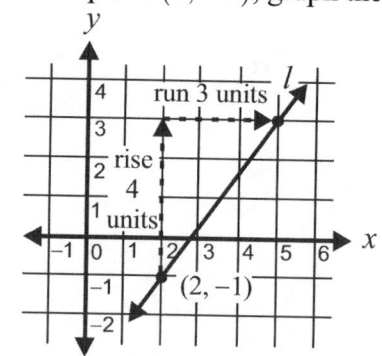

Draw the line through the two points.

Example 16: Given a line that has a slope of $-\frac{1}{4}$ and passes through the point $(-3, 2)$, graph the line.

Plot the point $(-3, 2)$.

Since the slope is negative, go **down** 1 unit and over 4 units to get a second point.

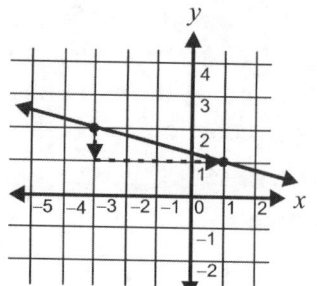

Graph the line through the two points.

Graph a line on your own graph paper for each of the following problems. First, plot the point. Then use the slope to find a second point. Draw the line formed from the point and the slope.

1. $(2, -2), m = \frac{3}{4}$
2. $(3, -4), m = \frac{1}{2}$
3. $(1, 3), m = -\frac{1}{3}$
4. $(2, -4), m = 1$
5. $(3, 0), m = -\frac{1}{2}$
6. $(-2, 1), m = \frac{4}{3}$
7. $(-4, -2), m = \frac{1}{2}$
8. $(1, -4), m = \frac{3}{4}$
9. $(2, -1), m = -\frac{1}{2}$
10. $(5, -2), m = \frac{1}{4}$
11. $(-2, -3), m = \frac{2}{3}$
12. $(4, -1), m = -\frac{1}{3}$
13. $(-1, 5), m = \frac{2}{5}$
14. $(-2, 3), m = \frac{3}{4}$
15. $(4, 4), m = -\frac{1}{2}$
16. $(3, -3), m = -\frac{3}{4}$
17. $(-2, 5), m = \frac{1}{3}$
18. $(-2, -3), m = -\frac{3}{4}$
19. $(4, -3), m = \frac{2}{3}$
20. $(1, 4), m = -\frac{1}{2}$

9.11 Finding the Equation of a Line Using Two Points or a Point and Slope

If you can find the slope of a line and know the coordinates of one point, you can write the equation for the line. You know the formula for the slope of a line is:

$$m = \frac{y_2 - y_1}{x_2 - x_1} \text{ or } \frac{y_2 - y_1}{x_2 - x_1} = m$$

Using algebra, you can see that if you multiply both sides of the equation by $x_2 - x_1$, you get:

$$y - y_1 = m(x - x_1) \longleftarrow \text{ point-slope form of an equation}$$

Example 17: Write the equation of the line passing through the points $(-2, 3)$ and $(1, 5)$.

Step 1: First, find the slope of the line using the two points given.
$$m = \frac{y_2 - y_1}{x_2 - x_1} = \frac{5 - 3}{1 - (-2)} = \frac{2}{3}$$

Step 2: Pick one of the two points to use in the point-slope equation. For point $(-2, 3)$, we know $x_1 = -2$ and $y_1 = 3$, and we know $m = \frac{2}{3}$. Substitute these values into the point-slope form of the equation.
$$y - y_1 = m(x - x_1)$$
$$y - 3 = \frac{2}{3}[x - (-2)]$$
$$y - 3 = \frac{2}{3}x + \frac{4}{3}$$
$$y = \frac{2}{3}x + \frac{13}{3}$$

Use the point-slope formula to write an equation for each of the following lines.

1. $(1, -2), m = 2$
2. $(-3, 3), m = \frac{1}{3}$
3. $(4, 2), m = \frac{1}{4}$
4. $(5, 0), m = 1$
5. $(3, -4), m = \frac{1}{2}$
6. $(-1, -4)\ (2, -1)$
7. $(2, 1)\ (-1, -3)$
8. $(-2, 5)\ (-4, 3)$
9. $(-4, 3)\ (2, -1)$
10. $(3, 1)\ (5, 5)$
11. $(-3, 1), m = 2$
12. $(-1, 2), m = \frac{4}{3}$
13. $(2, -5), m = -2$
14. $(-1, 3), m = \frac{1}{3}$
15. $(0, -2), m = -\frac{3}{2}$

9.12 Equations of Parallel Lines

If two linear equations have the same slope but different y-intercepts, they are **parallel** lines. Parallel lines never touch each other, so they have no points in common.

Example 18: Consider line l shown in Figure 2 at right. The equation of the line is $y = -\frac{1}{2}x + 3$. What happens to the graph of the line if the y-intercept is changed to -1?

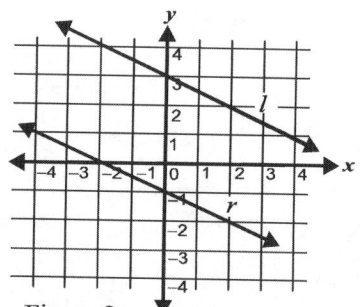

Figure 2

Rewrite the equation of the line replacing the y-intercept with -1. The equation of the new line is $y = -\frac{1}{2}x - 1$.

Graph the new line. Line r in Figure 2 is the graph of the equation $y = -\frac{1}{2}x - 1$. Since both lines l and r have the same slope, they are parallel. Line r, with a y-intercept of -1, sits below line l, with a y-intercept of 3.

Put each pair of the following equations in slope-intercept form. Write P if the lines are parallel and NP if the lines are not parallel.

1. $y = x + 1$
   $2y - 2x = 6$ _____

2. $2x + y = 6$
   $2x = 8 - y$ _____

3. $x + 5y = 0$
   $5y + 5 = x$ _____

4. $y = 3 - \frac{1}{3}x$
   $3y + x = -6$ _____

5. $x = 2y$
   $-x = -2y + 14$ _____

6. $y = x + 2$
   $-y = x + 4$ _____

7. $y = 4 - \frac{1}{4}x$
   $3x + 4y = 4$ _____

8. $x + y = 5$
   $5 - y = 2x$ _____

9. $x - 4y = 0$
   $4y = x - 8$ _____

9.13 Equations of Perpendicular Lines

Now that we know how to calculate the slope of lines using two points, we are going to learn how to calculate the slope of a line perpendicular to a given line, then find the equation of that perpendicular line. To find the slope of a line perpendicular to any given line, take the slope of the first line, m:

1. multiply the slope by -1
2. invert (or flip over) the slope

You now have the slope of a perpendicular line. Writing the equation for a line perpendicular to another line involves three steps:

1. find the slope of the perpendicular line
2. choose one point on the first line
3. use the point-slope form to write the equation

9.13 Equations of Perpendicular Lines

Example 19: The solid line on the graph below has a slope of $\frac{2}{3}$. Write the equation of a line perpendicular to the solid line.

Find the slope of the solid line. Multiply the slope by -1 and then find the inverse (flip it over).

$$\frac{2}{3} \times -1 = -\frac{2}{3} \curvearrowright -\frac{3}{2}$$

The slope of the perpendicular line, shown as a dotted line on the graph below, is $-\frac{3}{2}$.

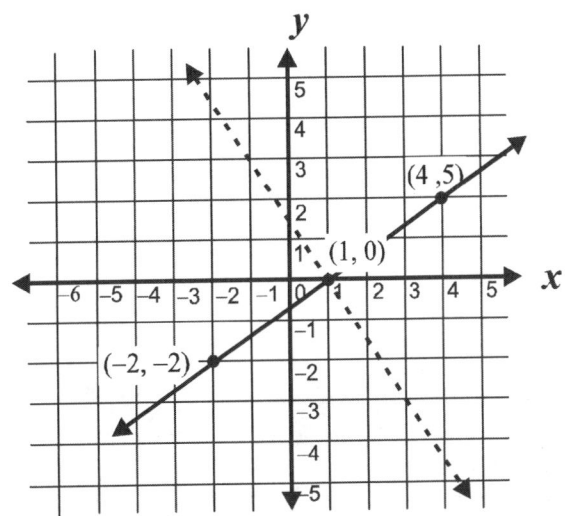

Step 2: Choose one point on the first line. We will use $(1, 0)$ in this example. The point $(-2, -2)$ or $(4, 5)$ could also be used.

Step 3: Use the point-slope formula, $(y - y_1) = m(x - x_1)$, to write the equation of the perpendicular line. Remember, we chose $(1, 0)$ as our point. So, $(y - 0) = -\frac{3}{2}(x - 1)$. Simplified, $y = -\frac{3}{2}x + \frac{3}{2}$.

Copyright © American Book Company

Chapter 9 Graphing and Writing Equations and Inequalities

Solve the following problems involving perpendicular lines.

1. Find the slope of the line perpendicular to the solid line shown at right, and draw the perpendicular as a dotted line. Use the point $(-1, 0)$ on the solid line and the calculated slope to find the equation of the perpendicular line.

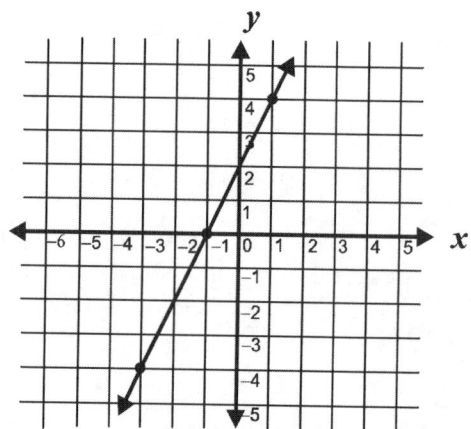

Find the equation of the perpendicular line using the point and slope given and the formula $(y - y_1) = m(x - x_1)$.

2. $(2, 1), 5$

3. $(3, 2), 2$

4. $(-2, 1), -3$

5. $(-4, 2), -\dfrac{1}{2}$

6. $(-1, 4), 1$

7. $(3, 3), \dfrac{2}{3}$

8. $(5, -1), -1$

9. $\left(\dfrac{1}{2}, \dfrac{3}{4}\right), 4$

10. $\left(\dfrac{2}{3}, \dfrac{3}{4}\right), -\dfrac{1}{6}$

11. $(7, -2), -\dfrac{1}{8}$

12. $(5, 0), \dfrac{4}{5}$

13. $(-3, -3), -\dfrac{7}{3}$

14. $\left(\dfrac{1}{4}, 4\right), \dfrac{1}{2}$

15. $(0, 6), -\dfrac{1}{9}$

9.14 Graphing Inequalities

In the previous section, you would graph the equation $x = 3$ as:

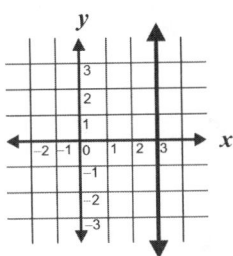

In this section, we graph inequalities such as $x > 3$ (read x is greater than 3). To show this, we use a broken line since the points on the line $x = 3$ are not included in the solution. We shade all points greater than 3.

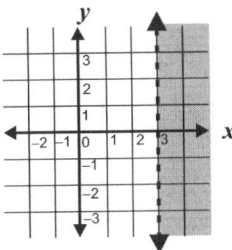

When we graph $x \geq 3$ (read x is greater than or equal to 3), we use a solid line because the points on the line $x = 3$ are included in the graph.

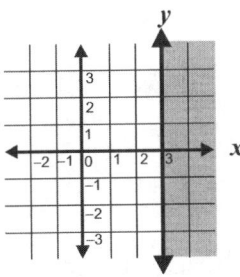

Graph the following inequalities on your own graph paper.

1. $y < 2$
2. $x \geq 4$
3. $y \geq 1$
4. $x < -1$
5. $y \geq -2$
6. $x \leq -4$
7. $x > -3$
8. $y \leq 3$
9. $x \leq 5$
10. $y > -5$
11. $x \geq 3$
12. $y < -1$
13. $x \leq 0$
14. $y > -1$
15. $y \leq 4$
16. $x \geq 0$
17. $y \geq 3$
18. $x < 4$
19. $x \leq -2$
20. $y < -2$
21. $y \geq -4$
22. $x \geq -1$
23. $y \leq 5$
24. $x < -3$

Chapter 9 Graphing and Writing Equations and Inequalities

Example 20: Graph $x + y \geq 3$.

Step 1: First, we graph $x + y \geq 3$ by changing the inequality to an equality. Think of ordered pairs that will satisfy the equation $x + y = 3$. Then, plot the points, and draw the line. As shown below, this line divides the Cartesian plane into 2 half-planes, $x + y \geq 3$ and $x + y \leq 3$. One half-plane is above the line, and the other is below the line.

x	y
2	1
0	3
3	0
4	-1

Step 2: To determine which side of the line to shade, first choose a test point. If the point you choose makes the inequality true, then the point is on the side you shade. If the point you choose does not make the inequality true, then shade the side that does not contain the test point.

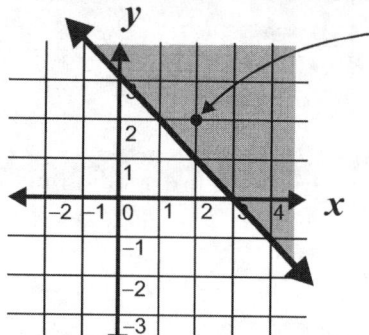

For our test point, let's choose $(2, 2)$. Substitute $(2, 2)$ into the inequality.

$$x + y \geq 3$$
$$2 + 2 \geq 3$$

$4 \geq 3$ is true, so shade the side that includes this point.

Use a solid line because of the $\geq$ sign.

Graph the following inequalities on your own graph paper.

1. $x + y \leq 4$
2. $x + y \geq 3$
3. $x \geq 5 - y$
4. $x \leq 1 + y$
5. $x - y \geq -2$
6. $x < y + 4$
7. $x + y < -1$
8. $x - y \leq 0$
9. $x \geq y + 2$
10. $x < -y + 1$
11. $-x + y > 1$
12. $-x - y < -2$

9.14 Graphing Inequalities

For more complex inequalities, it is easier to graph by first changing the inequality to an equality and then put the equation in slop-intercept form.

Example 21: Graph the inequality $2x + 4y \leq 8$.

Step 1: Change the inequality to an equality.
$2x + 4y = 8$

Step 2: Put the equation in slope-intercept form by solving the equation for y.

$$2x + 4y = 8$$
$$2x - 2x + 4y = -2x + 8 \quad \text{Subtract } 2x \text{ from both sides of the equation.}$$
$$4y = -2x + 8 \quad \text{Simplify.}$$
$$\frac{4y}{4} = \frac{-2x + 8}{4} \quad \text{Divide both sides by 4.}$$
$$y = \frac{-2x}{4} + \frac{8}{4} \quad \text{Find the lowest terms of the fractions.}$$
$$y = -\frac{1}{2}x + 2$$

Step 3: Graph the line. If the inequality is $<$ or $>$, use a dotted line. If the inequality is $\leq$ or $\geq$, use a solid line. For this example, we should use a solid line.

Step 4: Determine which side of the line to shade. Pick a point such as $(0,0)$ to see if it is true in the inequality.

$2x + 4y \leq 8$, so substitute $(0,0)$.
Is $0 + 0 \leq 8$? Yes, $0 \leq 8$, so shade the side of the line that includes the point $(0,0)$.

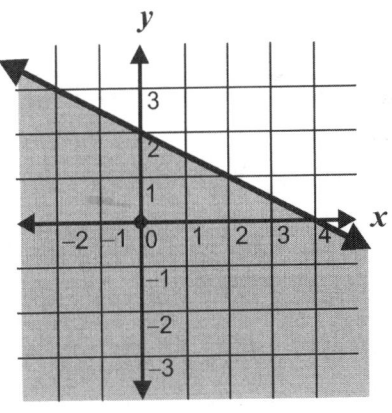

Graph the following inequalities on your own graph paper.

1. $2x + y \geq 1$
2. $3x - y \leq 3$
3. $x + 3y > 12$
4. $4x - 3y < 12$
5. $y \geq 3x + 1$
6. $x - 2y > -2$
7. $x \leq y + 4$
8. $x + y < -1$
9. $-4y \geq 2x + 1$
10. $x \leq 4y - 2$
11. $3x - y \geq 4$
12. $y \geq 2x - 5$

Chapter 9 Review

1. Graph the solution set for the linear equation: $x - 3 = y$.

2. Graph the equation $2x - 4 = 0$.

3. What is the slope of the line that passes through the points $(5, 3)$ and $(6, 1)$?

4. What is the slope of the line that passes through the points $(-1, 4)$ and $(-6, -2)$?

5. What is the x-intercept for the following equation? $6x - y = 30$

6. What is the y-intercept for the following equation? $4x + 2y = 28$

7. Graph the equation $3y = 9$.

8. Write the following equation in slope-intercept form.
$3x = -2y + 4$

9. What is the slope of the line $y = -\frac{1}{2}x + 3$?

10. What is the x-intercept of the line $y = 5x + 6$?

11. What is the y-intercept of the line $y - \frac{2}{3}x + 3 = 0$?

12. Graph the line which has a slope of -2 and a y-intercept of -3.

13. Find the equation of the line which contains the point $(0, 2)$ and has a slope of $\frac{3}{4}$.

14. What is the distance between the points $(3, 3)$ and $(6, -1)$?

15. What is the distance between the two points $(-3, 0)$ and $(2, 5)$?

For questions 16 and 17, use the following formula to find the coordinates of the midpoint of the line segments with the given endpoints.

$$\text{midpoint} = \left(\frac{x_1 + x_2}{2}, \frac{y_1 + y_2}{2}\right)$$

16. $(6, 10)$ $(-4, 4)$

17. $(-1, -7)$ $(5, 3)$

Graph the following inequalities on a Cartesian plane using your graph paper.

18. $x \geq 4$

19. $x \leq -2$

20. $5y > -10x + 5$

21. $y \leq 2$

22. $2x + y < 5$

23. $y - 2x \leq 3$

24. $y \geq x + 2$

25. $3 + y > x$

Chapter 9 Test

1. Which of the following is not a solution of $3x = 5y - 1$?

   (A) $(3, 2)$
   (B) $(7, 4)$
   (C) $\left(-\frac{1}{3}, 0\right)$
   (D) $(-2, -1)$

2. $(-2, 1)$ is a solution for which of the following equations?

   (A) $y + 2x = 4$
   (B) $-2x - y = 5$
   (C) $x + 2y = -4$
   (D) $2x - y = -5$

3. Which is the graph of $x - 3y = 6$?

   (A)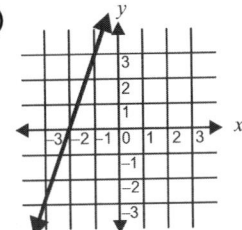

   (B)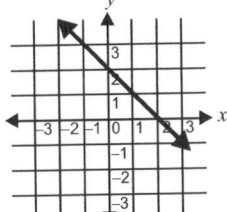

   (C)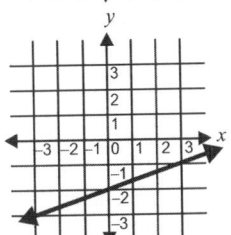

   (D)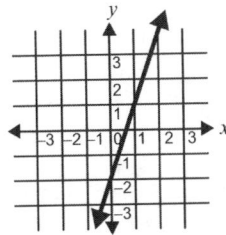

4. Which of the following points does **not** lie on the line $y = 3x - 2$?

   (A) $(0, -2)$
   (B) $(1, 1)$
   (C) $(-1, 5)$
   (D) $(2, 4)$

5. Which of the following is the graph of the equation $y = x - 3$?

   (A)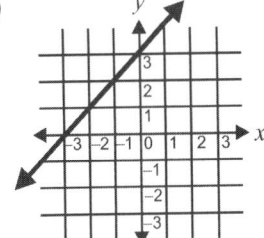

   (B)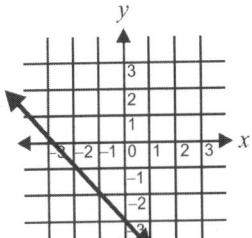

   (C)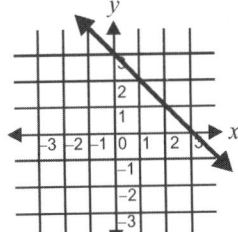

   (D)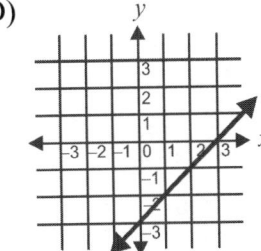

Chapter 9 Graphing and Writing Equations and Inequalities

6. What is the x-intercept of the following linear equation? $3x + 4y = 12$
   (A) $(0, 3)$
   (B) $(3, 0)$
   (C) $(0, 4)$
   (D) $(4, 0)$

7. Which of the following equations is represented by the graph?

   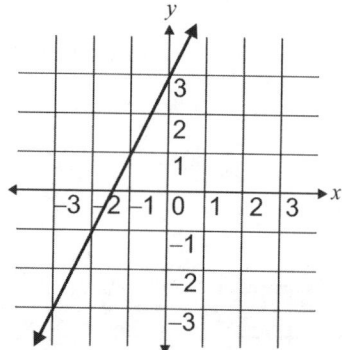

   (A) $y = -3x + 3$
   (B) $y = -\frac{1}{3}x + 3$
   (C) $y = 3x - 3$
   (D) $y = 2x + 3$

8. What is the equation of the line that includes the point $(4, -3)$ and has a slope of -2?
   (A) $y = -2x - 5$
   (B) $y = -2x - 2$
   (C) $y = -2x + 5$
   (D) $y = 2x - 5$

9. What is the x-intercept and y-intercept for the equation $x + 2y = 6$?
   (A) x-intercept $= (0, 6)$
       y-intercept $= (3, 0)$
   (B) x-intercept $= (4, 1)$
       y-intercept $= (2, 2)$
   (C) x-intercept $= (0, 6)$
       y-intercept $= (0, 3)$
   (D) x-intercept $= (6, 0)$
       y-intercept $= (0, 3)$

10. Put the following equation is slope-intercept form.

    $2x - 3y = 6$

    (A) $y = \frac{2}{3}x - 2$
    (B) $y = 2x - 2$
    (C) $y = -\frac{2}{3}x + 2$
    (D) $y = 2x + 2$

11. Which of the following is a graph of the inequality $-y \geq 2$?

    (A)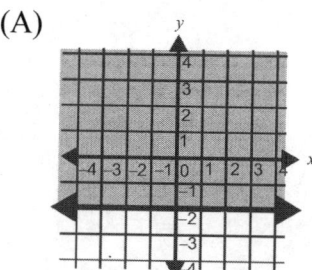

    (B)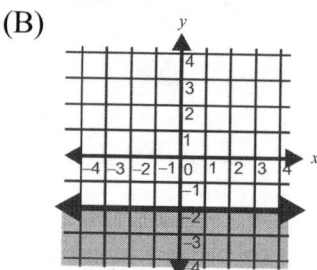

    (C)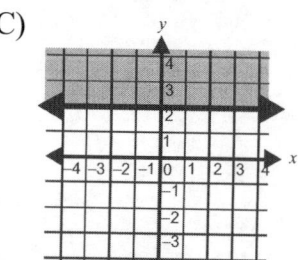

    (D)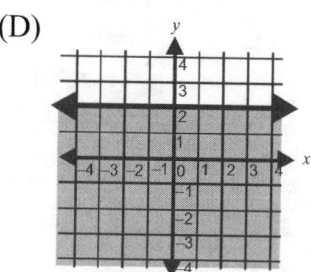

12. Which of the following graphs shows a line with a slope of 0 that passes through the point $(3, 2)$?

(A)

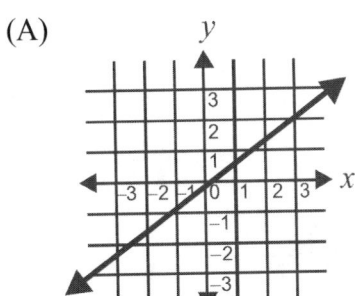

(B)

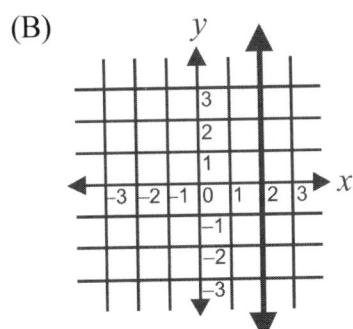

(C)

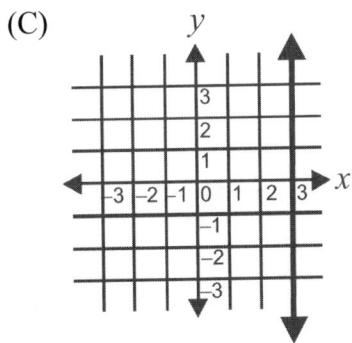

(D)

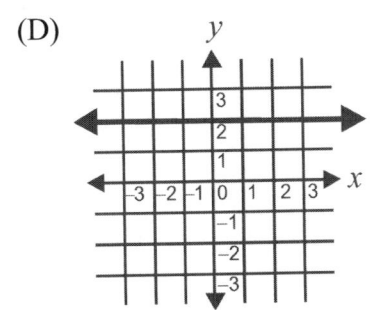

13. Which of the following is a graph of the inequality $y \leq x - 3$?

(A)

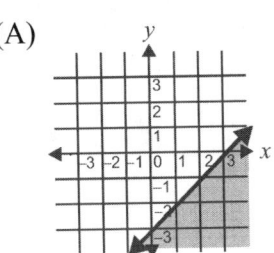

(B)

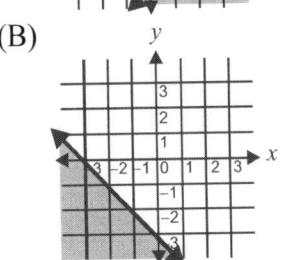

(C)

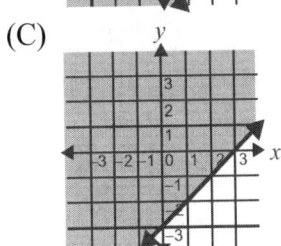

(D)

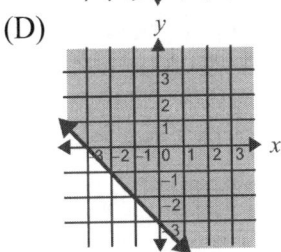

14. The graph of which pair of equations below will be parallel?

(A) $x + 4y = 3$
$3x + 4y = 3$

(B) $x - 4y = 3$
$4y - x = -3$

(C) $2x - 8 = 2y$
$2x + 8 = 2y$

(D) $6x + 6 = 6y$
$11x - 12 = 7y$

Chapter 9 Graphing and Writing Equations and Inequalities

15. Look at the graphs below. Which of the following statements is false?

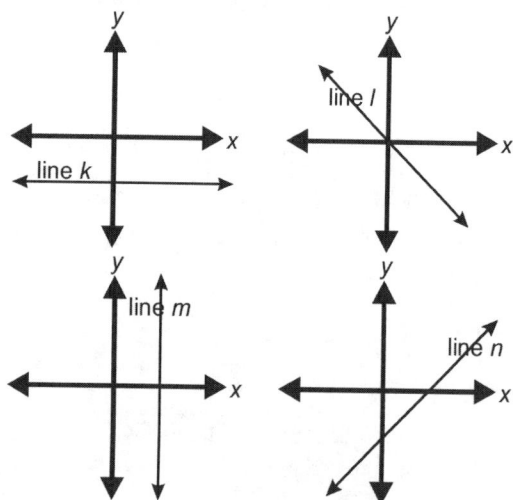

(A) The slope of line k is undefined.
(B) The slope of line l is negative.
(C) The slope of line m is undefined.
(D) The slope of line n is positive.

16. The coordinates of a line segment are $(-5, 3)$ and $(3, -7)$. What are the coordinates for the midpoint?

(A) $(4, -2)$
(B) $(4, 5)$
(C) $(-1, -2)$
(D) $(-4, -5)$

17. The coordinates of a line segment are $(1, 6)$ and $(11, -4)$. What are the coordinates for the midpoint?

(A) $(6, 1)$
(B) $(10, 2)$
(C) $(5, 1)$
(D) $(12, 2)$

18. Which of the following is an equation of a line that is perpendicular to the line l in the graph?

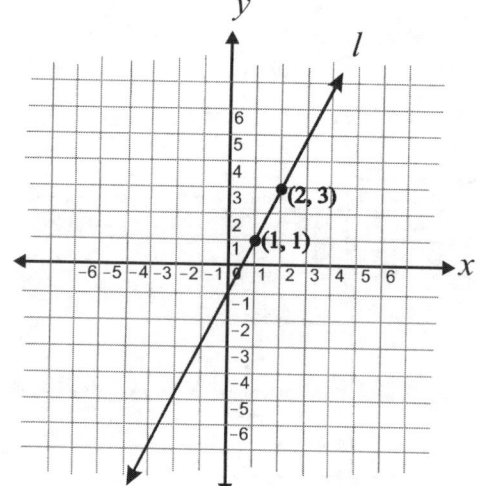

(A) $x - 2y = -4$
(B) $x - 2y = 4$
(C) $x + 2y = 4$
(D) $2x + y = 4$

Chapter 10
Mathematical Reasoning

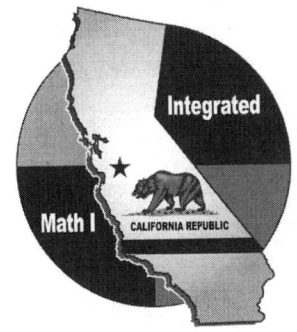

This chapter covers the following CA Integrated Math I standards:

Algebra	25.1
	25.3

10.1 Mathematical Reasoning/Logic

The CA Integrated Math I test calls for skill development in mathematical **reasoning** or **logic**. The ability to use logic is an important skill for solving math problems, but it can also be helpful in real-life situations. For example, if you need to get to Park Street, and the Park Street bus always comes to the bus stop at 3 PM, then you know that you need to get to the bus stop by at least 3 PM. This is a real-life example of using logic, which many people would call "common sense."

There are many different types of statements which are commonly used to describe mathematical principles. However, using the rules of logic, the truth of any mathematical statement must be evaluated. Below is a list of tools used in logic to evaluate mathematical statements.

Logic is the discipline that studies valid reasoning. There are many forms of valid arguments, but we will just review a few here.

A **proposition** is usually a declarative sentence which may be true or false.

An **argument** is a set of two or more related propositions, called **premises**, that provide support for another proposition, called the **conclusion**.

Deductive reasoning is an argument which begins with general premises and proceeds to a more specific conclusion. Most elementary mathematical problems use deductive reasoning.

Inductive reasoning is an argument in which the truth of its premises make it likely or probable that its conclusion is true.

Chapter 10   Mathematical Reasoning

ARGUMENTS

Most of logic deals with the evaluation of the validity of arguments. An argument is a group of statements that includes a conclusion and at least one premise. A premise is a statement that you know is true or at least you assume to be true. Then, you draw a conclusion based on what you know or believe is true in the premise(s). Consider the following example:

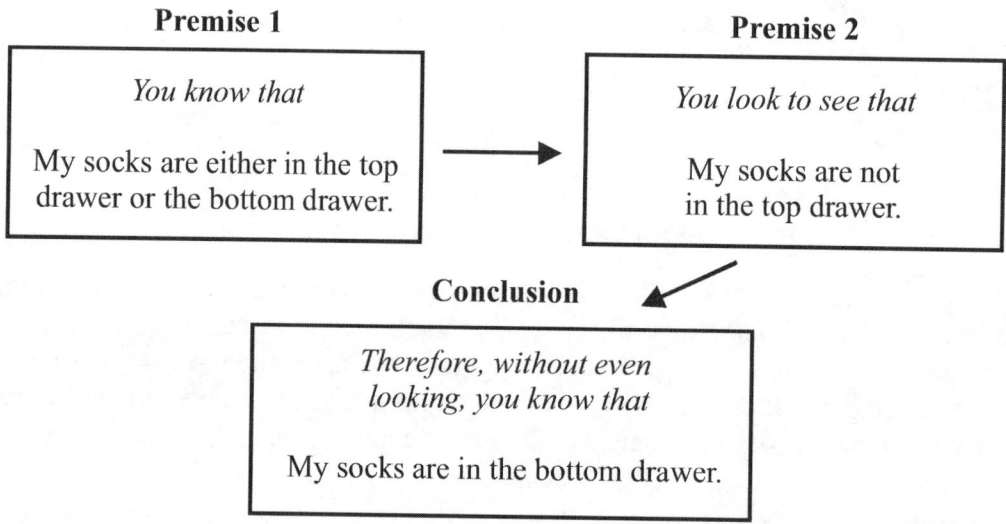

This argument is an example of deductive reasoning, where the conclusion is "deduced" from the premises and nothing else. In other words, if Premise 1 and Premise 2 are true, you don't even need to look in the bottom drawer to know that the conclusion is true.

For numbers 1–5, what conclusion can be drawn from each proposition?

1. All squirrels are rodents. All rodents are mammals. Therefore,

2. All fractions are rational numbers. All rational numbers are real numbers. Therefore,

3. All squares are rectangles. All rectangles are parallelograms. All parallelograms are quadrilaterals. Therefore,

4. All Chevrolets are made by General Motors. All Luminas are Chevrolets. Therefore,

5. If a number is even and divisible by three, then it is divisible by six. Eighteen is divisible by six. Therefore,

10.2 Deductive and Inductive Arguments

In general, there are two types of logical arguments: **deductive** and **inductive**. Deductive arguments tend to move from general statements or theories to more specific conclusions. Inductive arguments tend to move from specific observations to general theories.

Deductive Reasoning

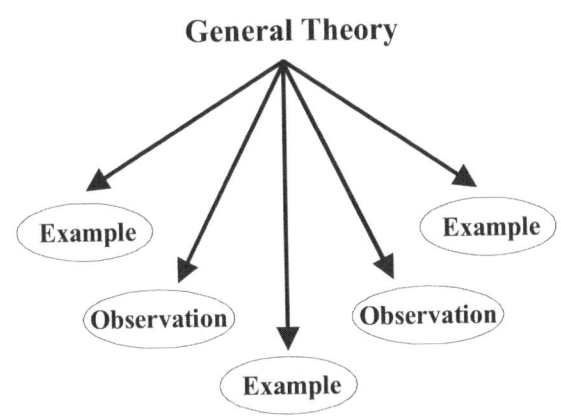

Inductive Reasoning

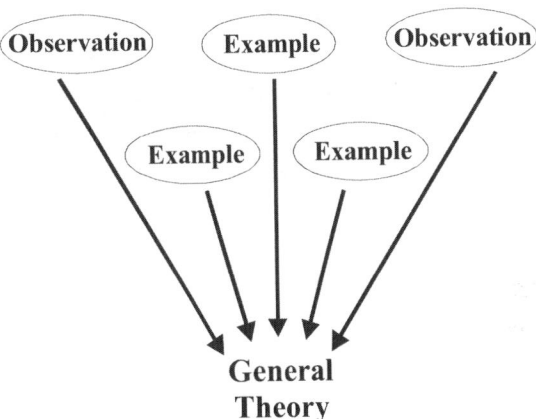

Compare the two examples below:

Deductive Argument

Premise 1	All men are mortal.
Premise 2	Socrates is a man.
Conclusion	Socrates is mortal.

Inductive Argument

Premise 1	The sun rose this morning.
Premise 2	The sun rose yesterday morning.
Premise 3	The sun rose two days ago.
Premise 4	The sun rose three days ago.
Conclusion	The sun will rise tomorrow.

An inductive argument cannot be proved beyond a shadow of a doubt. For example, it's a pretty good bet that the sun will come up tomorrow, but the sun not coming up presents no logical contradiction.

On the other hand, a deductive argument can have logical certainty, but it must be properly constructed. Consider the examples below.

True Conclusion from an Invalid Argument	False Conclusion from a Valid Argument
All men are mortal. Socrates is mortal. Therefore Socrates is a man.	All astronauts are men. Julia Roberts is an astronaut. Therefore, Julia Roberts is a man.
Even though the above conclusion is true, the argument is based on invalid logic. Both men and women are mortal. Therefore, Socrates could be a woman.	In this case, the conclusion is false because the premises are false. However, the logic of the argument is valid because *if* the premises were true, then the conclusion would be true.

Chapter 10   Mathematical Reasoning

A **counterexample** is an example given in which the statement is true but the conclusion is false when we have assumed it to be true. If we said "All cocker spaniels have blonde hair," then a counterexample would be a red-haired cocker spaniel. If we made the statement, "If a number is greater than 10, it is less than 20," we can easily think of a counterexample, like 35.

Example 1:   Which argument is valid?

If you speed on Hill Street, you will get a ticket.
If you get a ticket, you will pay a fine.

(A) I paid a fine, so I was speeding on Hill Street.
(B) I got a ticket, so I was speeding on Hill Street.
(C) I exceeded the speed limit on Hill Street, so I paid a fine.
(D) I did not speed in Hill Street, so I did not pay a fine.

Solution: C is valid.
A is incorrect. I could have paid a fine for another violation.
B is incorrect. I could have gotten a ticket for some other violation.
D is incorrect. I could have paid a fine for speeding somewhere else.

Example 2:   Assume the given proposition is true. Then, determine if each statement is true or false.

Given: If a dog is thirsty, he will drink.

(A)   If a dog drinks, then he is thirsty.   T or F
(B)   If a dog is not thirsty, he will not drink.   T or F
(C)   If a dog will not drink, he is not thirsty.   T or F

Solution: A is false. He is not necessarily thirsty; he could just drink because other dogs are drinking or drink to show others his control of the water. This statement is the **converse** of the original. The converse of the statement "If A, then B" is "If B, then A."

B is false. The reasoning from A applies. This statement is the **inverse** of the original. The inverse of the statement "If A, then B" is "If not A, then not B."

C is true. It is the **contrapositive**, or the complete opposite of the original. The contrapositive says "If not B, then not A."

For numbers 1–4, assume the given proposition is true. Then, determine if the statements following it are true or false.

All squares are rectangles.

1. All rectangles are squares.   T or F
2. All non-squares are non-rectangles.   T or F
3. No squares are non-rectangles.   T or F
4. All non-rectangles are non-squares.   T or F

Chapter 10 Review

For numbers 1–4, assume the given proposition is true. Then determine if the statements following it are true or false.

All whales are mammals.

1. All non-whales are non-mammals. T or F
2. If a mammal lives in the sea, it is a whale. T or F
3. All mammals are whales. T or F
4. All non-mammals are non-whales. T or F

For numbers 5–8, determine whether the situation is showing deductive or inductive logic.

5. A group of students were given three descriptions about a person's job. They were then told to decide what type of job title the person has.

6. When traveling in a car on a family vacation, I noticed that I could see the ocean to my left and palm trees to my right. I concluded that my family and I were going to the beach.

7. Sammy asked her friend, Amy, to give her a good reason to get a summer job. Amy gave Sammy four good reasons to get a job.

8. The neighbor's cars are in the driveway and all of the lights in the house are off, so they must be sleeping.

Look at statements 9–12. Determine whether the statements are true always, sometimes, or never.

9. Quadratic equations have two solutions.

10. If you graph a linear equation, the graph will be a straight line.

11. When multiplying both sides of an inequality by a number, you must reverse the direction of the inequality symbol.

12. When you take the absolute value of a number, you are making the number negative.

Chapter 10   Mathematical Reasoning

Chapter 10 Test

For 1–3, chose which argument is valid.

1. If I oversleep, I miss breakfast. If I miss breakfast, I cannot concentrate in class. If I do not concentrate in class, I make bad grades.
   (A) I made bad grades today, so I missed breakfast.
   (B) I made good grades today, so I got up on time.
   (C) I could not concentrate in class today, so I overslept.
   (D) I had no breakfast today, so I overslept.

2. If I do not maintain my car regularly, it will develop problems. If my car develops problems, it will not be safe to drive. If my car is not safe to drive, I cannot take a trip in it.
   (A) If my car develops problems, I did not maintain it regularly.
   (B) I took a trip in my car, so I maintained it regularly.
   (C) If I maintain my car regularly, it will not develop problems.
   (D) If my car is safe to drive, it will not develop problems.

3. If two triangles have all corresponding sides and all corresponding angles congruent, then they are congruent triangles. If two triangles are congruent, then they are similar triangles.
   (A) Similar triangles have all sides and all angles congruent.
   (B) If two triangles are similar, then they are congruent.
   (C) If two triangles are not congruent, then they are not similar.
   (D) If two triangles have all corresponding sides and angles congruent, then they are similar triangles.

4. Cynthia is asked to list five duties of the President. What type of logic is Cynthia using?
   (A) mathematical reasoning
   (B) inductive reasoning
   (C) intuitive reasoning
   (D) deductive reasoning

5. You can find the equation of a line by using two points that lie on that line. When is this statement true?
   (A) always
   (B) sometimes
   (C) never
   (D) cannot be determined

6. When using order of operations to simplify an algebraic expression, the first step is to get rid of the exponents. When is this statement true?
   (A) always
   (B) sometimes
   (C) never
   (D) cannot be determined

Chapter 11
Triangles

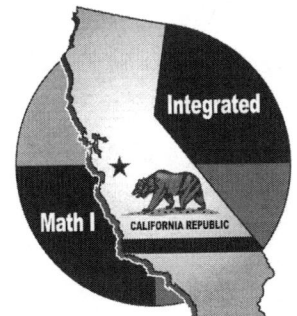

This chapter covers the following CA Integrated Math I standards:

Geometry	6.0
	12.0
	18.0

11.1 Types of Triangles

right triangle
contains 1 right ∠

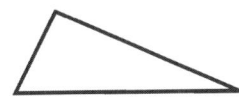

acute triangle
all angles are acute
(less than 90°)

obtuse triangle
one angle is obtuse
(greater than 90°)

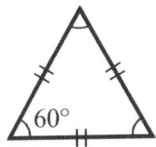

equilateral triangle
all three sides equal
all angles are 60°

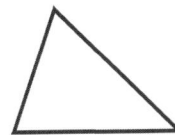

scalene triangle
no sides equal
no angles equal

isosceles triangle
two sides equal
two angles equal

A triangle is an **equilateral triangle** if all of its sides and angles are equal. A triangle is an **isosceles triangle** if two of its sides and the angles opposite those sides are equal. A triangle is a **right triangle** if one of its angles equals 90°.

11.2 Interior Angles of a Triangle

The three interior angles of a triangle always add up to 180°.

Example 1:

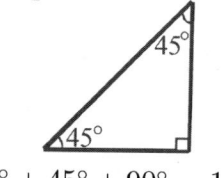
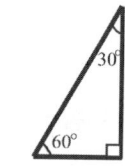
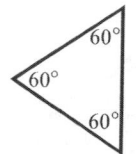

$45° + 45° + 90° = 180°$ $30° + 60° + 90° = 180°$ $60° + 60° + 60° = 180°$

Example 2: Find the missing angle in the triangle.

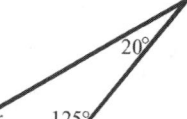

Solution:

$$20° + 125° + x = 180°$$
$$-20° - 125° \quad\quad -20° - 125°$$
$$x = 180° - 20° - 125°$$
$$x = 35°$$

Subtract 20° and 125° from both sides to get x by itself.

The missing angle is 35°.

Find the missing angle in the triangles.

1.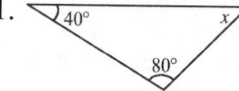

2.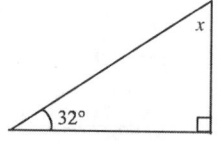

3.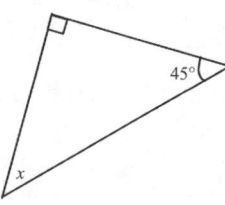

4.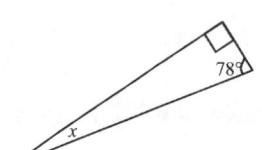

5.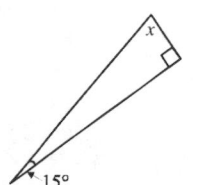

6.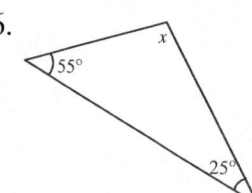

7.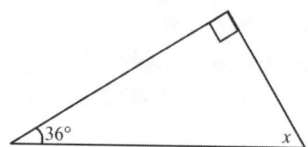

8.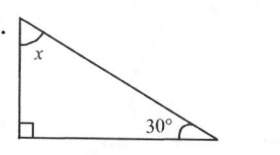

9.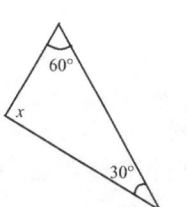

Find the missing angles in the triangles.

10.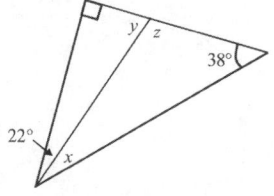

11.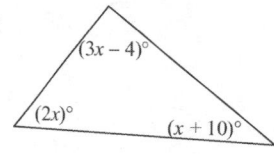

12.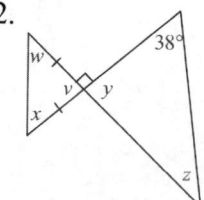

11.3 Exterior Angles

The **exterior angle** of a triangle is always equal to the sum of the opposite interior angles.

Example 3: Find the measure of $\angle x$ and $\angle y$.

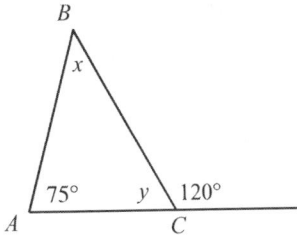

Step 1: Using the rule for exterior angles,
$120° = \angle A + \angle B$
$120° = 75° + x$
$120° - 75° = 75° - 75° + x$
$45° = x$

Step 2: The sum of the interior angles of a triangle equals 180°, so
$180° = 75° + 45° + y$
$180° - 75° - 45° = 75° - 75° + 45° - 45° + y$
$60° = y$

Find the measures of x and y.

1.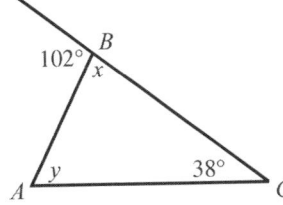

3.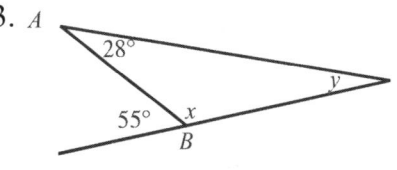

5.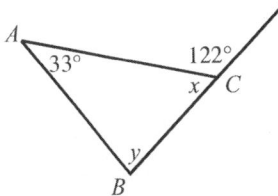

2.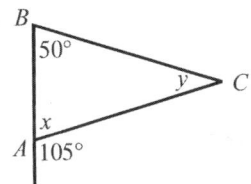

4.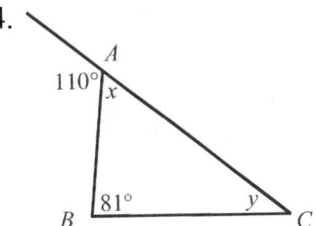

6.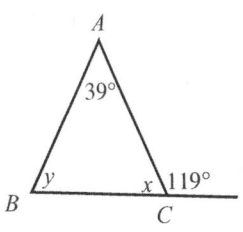

Find the measures of the angles.

7.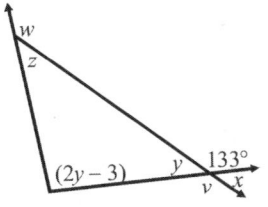

8.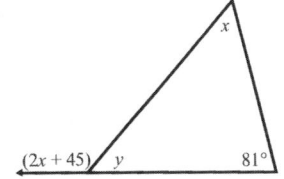

9.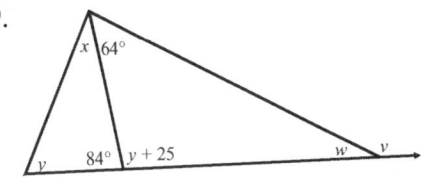

Chapter 11   Triangles

11.4   Triangle Inequality Theorem

The triangle inequality theorem states that the sum of the measure of any two sides in a triangle must be greater than the measure of the third side.

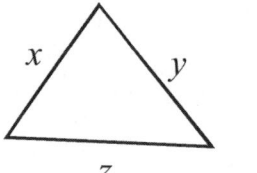

$x + y > z$
$y + z > x$
$x + z > y$

Example 4: Determine whether or not it is possible to create a triangle with sides of 1 unit, 5 units, and 7 units.

Step 1: First, you must set up three inequalities. Remember the sum of any two sides of a triangle must be greater than the third side.

$1 + 5 > 7$ $1 + 7 > 5$ $5 + 7 > 1$

Step 2: Determine if the inequalities are true.

$1 + 5 > 7$ $1 + 7 > 5$ $5 + 7 > 1$
$6 > 7$ $8 > 5$ $12 > 1$
False   True   True

The number 6 is not greater than 7, so a triangle cannot be formed using the sides given.
(All three inequalities must be true in order to create a triangle.)

Determine whether or not it is possible to create a triangle given the following measures of sides. Write yes if it is possible to form a triangle with the given measures of sides or write no if it is not possible.

1. 7, 8, 13
2. 2, 5, 9
3. 10, 8, 15
4. 6, 9, 20
5. 101, 89, 150
6. 1, 2, 4
7. 7, 7, 14
8. 21, 15, 29
9. 11, 9, 17

Chapter 11 Review

1. Find the missing angle.

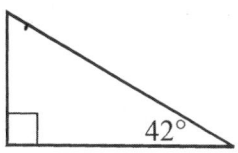

For questions 2–5, find the missing angles.

2.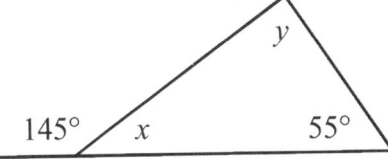

3.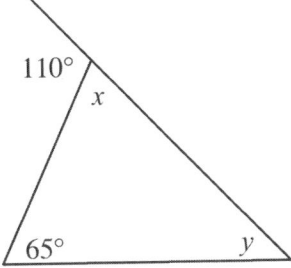

4.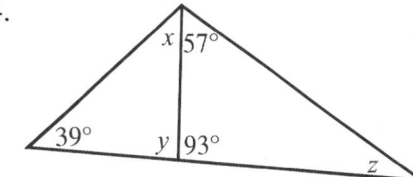

5.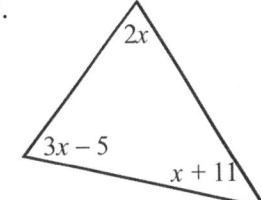

For questions 6–9, determine whether or not the measures of sides given can form a triangle using the triangle inequality theorem. Write yes if a triangle can be formed or write no if a triangle cannot be formed.

6. 1, 5, 3

7. 16, 22, 31

8. 19, 8, 10

9. 2, 2, 4

Chapter 11 Test

1. What is the measure of missing angle?

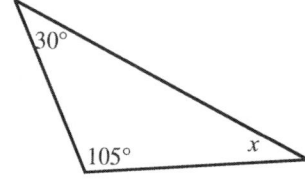

   (A) 225°
   (B) 45°
   (C) 75°
   (D) 30°

2. What is the measure of y?

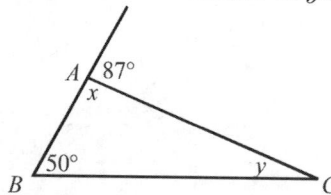

   (A) 93°
   (B) 87°
   (C) 37°
   (D) Cannot be determined

3. What is the measure of the two missing angles?

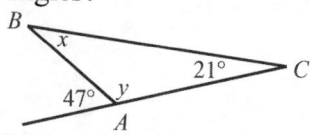

   (A) $x = 26°, y = 133°$
   (B) $x = 47°, y = 112°$
   (C) $x = 112°, y = 47°$
   (D) $x = 133°, y = 26°$

4. Given the measures of sides below, which cannot form a triangle?

   (A) 5, 7, 10
   (B) 2, 3, 4
   (C) 15, 6, 9
   (D) 19, 20, 36

5. Given the measures of sides below, which can form a triangle?

   (A) 2, 5, 9
   (B) 13, 6, 21
   (C) 4, 12, 16
   (D) 42, 5, 44

6. What is the measure of x in the triangle below?

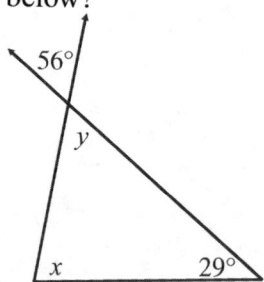

   (A) 56°
   (B) 95°
   (C) 124°
   (D) 85°

7. What is the measure of $\angle A$ in the figure below?

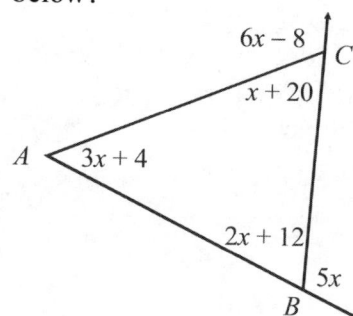

   (A) 24°
   (B) 44°
   (C) 60°
   (D) 76°

Chapter 12
Plane Geometry

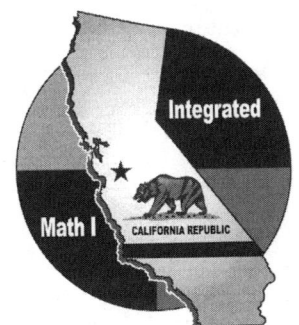

This chapter covers the following CA Integrated Math I standards:

Geometry	8.0
	10.0
	11.0
	12.0

12.1 Types of Polygons

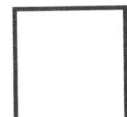

square
equal sides
90° ∠s

rectangle
opposite sides
parallel, 90° ∠s

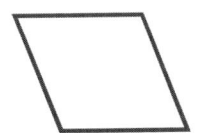

parallelogram
opposite sides
parallel

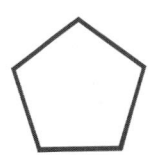

pentagon
5 sides

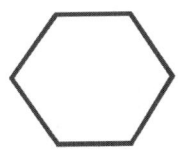

hexagon
6 sides

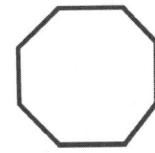

octagon
8 sides

12.2 Sum of Interior Angles of a Polygon

Given a polygon, you can find the sum of the measures of the interior angles using the following formula: Sum of the measures of the interior angles $= 180°(n-2)$, where n is the number of sides of the polygon.

Example 1: Find the sum of the measures of the interior angles of the following polygon:

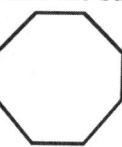

Solution: The figure has 8 sides. Using the formula we have $180°(8-2) = 180°(6) = 1080°$

Using the formula, $180°(n-2)$, find the sum of the interior angles of the following figures.

1.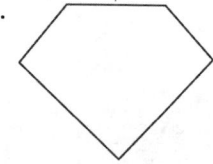
4.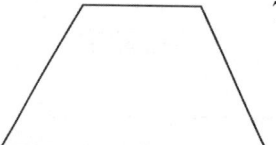
7.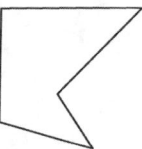
10.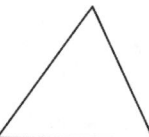

2.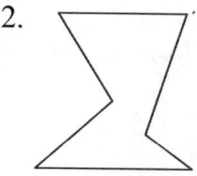
5.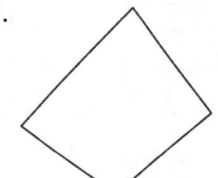
8.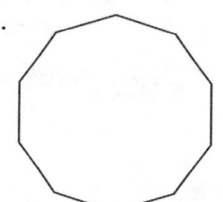
11.

3.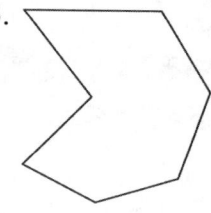
6.
9.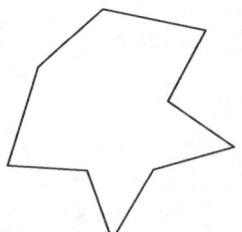
12.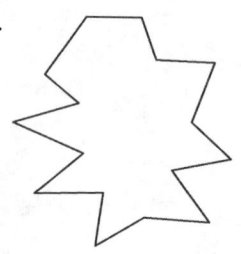

Find the measure of $\angle G$ in the regular polygons shown below. Remember that the sides of a regular polygon are equal.

13.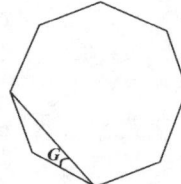
14.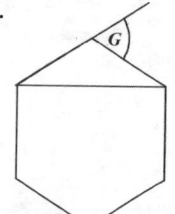
15.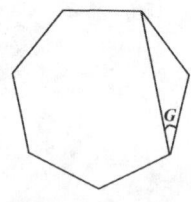

12.3 Perimeter

The **perimeter** is the distance around a polygon. To find the perimeter, add the lengths of the sides.
Examples:

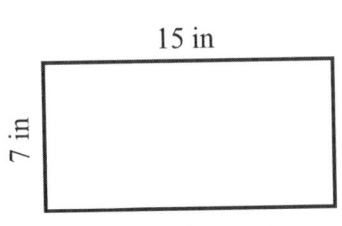

$P = 7 + 15 + 7 + 15$
$P = 44\,\text{in}$

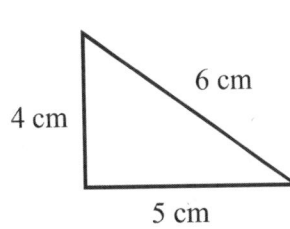

$P = 4 + 6 + 5$
$P = 15\,\text{cm}$

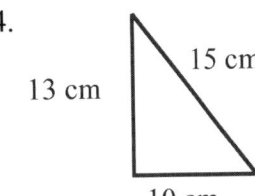

$P = 8 + 15 + 20 + 12 + 10$
$P = 65\,\text{ft}$

Find the perimeter of the following polygons.

1.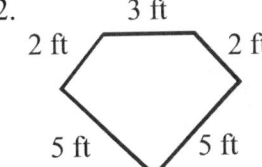

2.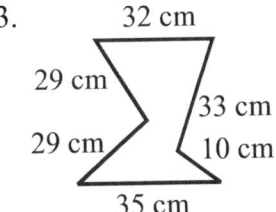

3. 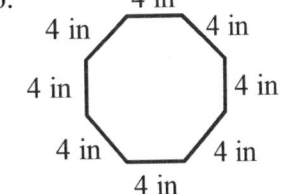

4.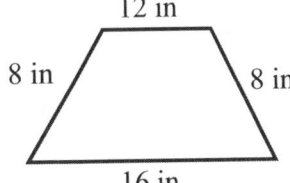

5.

6.

Chapter 12 Plane Geometry

12.4 Area of Squares and Rectangles

Area - area is always expressed in square units, such as in^2, m^2, and ft^2.

The area, (A), of squares and rectangles equals length (l) times width (w). $A = l \times w$.

Example 2:

4 cm

4 cm

$A = lw$
$A = 4 \times 4$
$A = 16$ **cm²**

If a square has an area of 16 cm^2**, it means that it will take** 16 **squares that are** 1 **cm on each side to cover the area that is** 4 **cm on each side.**

Find the area of the following squares and rectangles using the formula $A = lw$.

1. 10 ft
    10 ft

2. 5 cm
    2 cm

3. 4 in
    9 in

4. 9 in
    20 in

5. 6 ft
    6 ft

6. 10 cm
    5 cm

7. 4 ft
    2 ft

8. 5 in
    8 in

9. 12 ft
    12 ft

10. 7 cm
    12 cm

11. 1 ft
    8 ft

12. 6 cm
    7 cm

12.5 Area of Triangles

Example 3: Find the area of the following triangle.
The formula for the area of a triangle is as follows:

$$A = \frac{1}{2} \times b \times h$$

A = area
b = base
h = height or altitude

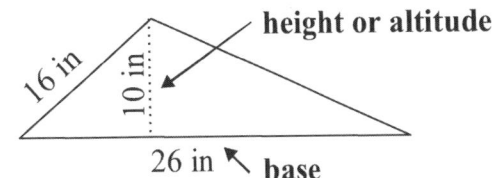

Step 1: Insert the measurements from the triangle into the formula: $A = \frac{1}{2} \times 26 \times 10$

Step 2: Cancel and multiply.
$$A = \frac{1}{\cancel{26}_{1}} \times \frac{\cancel{26}^{13}}{1} \times \frac{10}{1} = 130 \text{ in}^2$$

Note: Area is always expressed in square units such as in^2, ft^2, or m^2.

Find the area of the following triangles. Remember to include units.

1. 3 in, 4 in, 5 in
2. 12 cm, 6 cm height, 7 cm
3. 6 ft, 9 ft, 7 ft
4. 12 cm, 12 cm
5. 3 ft, 2 ft
6. 20 cm, 16 cm
7. 8 m height, 7 m, 15 m
8. 7 in, 9 in, 9 in
9. 2 ft, 2 ft
10. 5 ft, 4 ft, 6 ft
11. 12 ft, 10 ft, 15 ft
12. 3 m, 5 m, 10 m

Chapter 12 Plane Geometry

12.6 Area of Trapezoids and Parallelograms

Example 4:   Find the area of the following parallelogram.

The formula for the area of a parallelogram is $A = bh$.
A = area
b = base
h = height

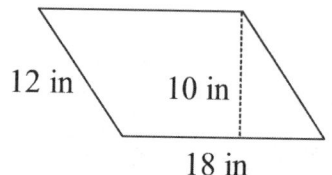

Step 1:   Insert measurements from the parallelogram into the formula: $A = 18 \times 10$.
Step 2:   Multiply. $18 \times 10 = 180$ in²

Example 5:   Find the area of the following trapezoid.
The formula for the area of a trapezoid is $A = \frac{1}{2}h(b_1 + b_2)$. A trapezoid has two bases that are parallel to each other. When you add the length of the two bases together and then multiply by $\frac{1}{2}$, you find their average length.

A = area
b = base
h = height

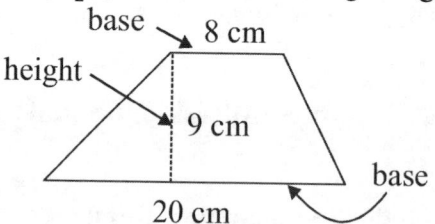

Insert the measurements from the trapezoid into the formula and solve:
$\frac{1}{2} \times 9 (8 + 20) = 126$ cm²

Find the area of the following parallelograms and trapezoids.

1.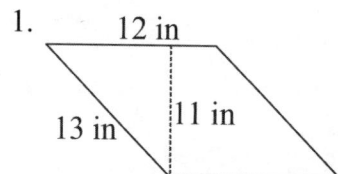

4.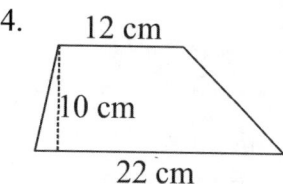

7.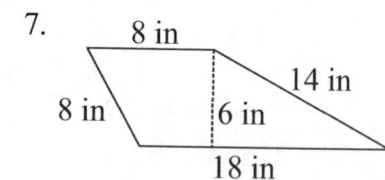

2.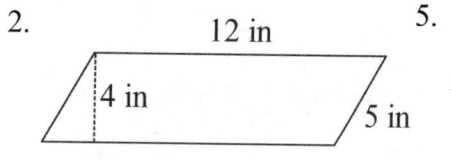

5.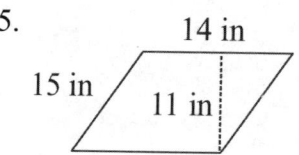

8.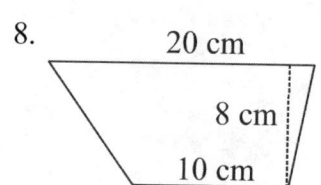

3.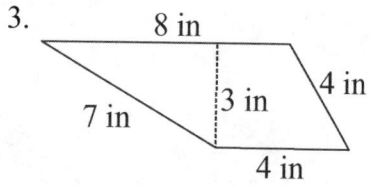

6.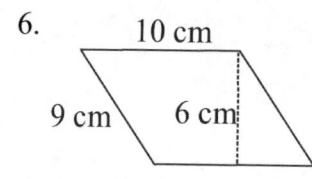

9.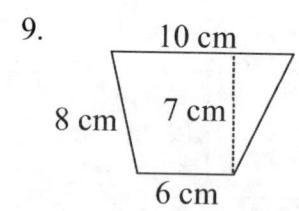

12.7 Area of a Rhombus

Example 6: Find the area of the following rhombus.

The formula for the area of a rhombus is as follows:

$$A = \frac{1}{2} \times d_1 \times d_2$$

A = area
d = diagonal

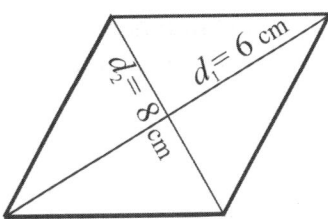

Step 1: Insert the measurements from the rhombus into the formula: $A = \frac{1}{2} \times 6 \times 8$

Step 2: Cancel and multiply. $A = \frac{1}{\underset{1}{2}} \times \frac{\overset{3}{\cancel{6}}}{1} \times \frac{8}{1} = 24 \text{ cm}^2$

Find the area of the following figures. Remember to include units.

1.

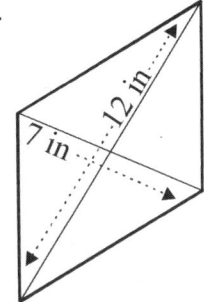

2.

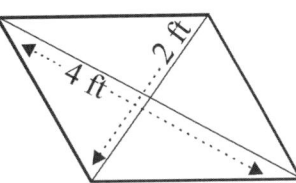

3.

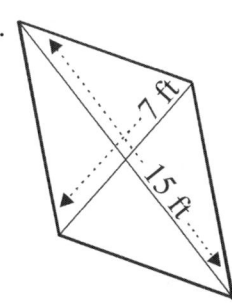

4.

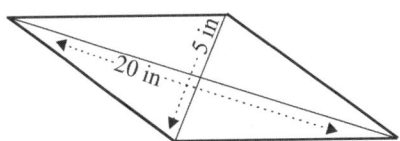

5.

6.

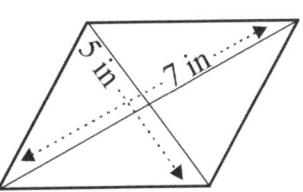

7.

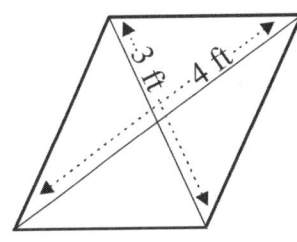

8.

9.

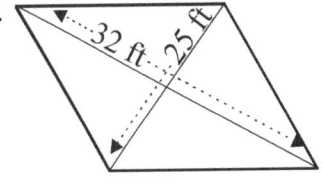

10.

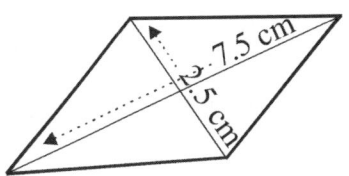

Chapter 12   Plane Geometry

12.8   Circumference

Circumference, C - the distance around the outside of a circle
Diameter, d - a line segment passing through the center of a circle from one side to the other
Radius, r - a line segment from the center of a circle to the edge of a circle
Pi, π - the ratio of a circumference of a circle to its diameter $\pi \approx 3.14$ or $\pi \approx \frac{22}{7}$

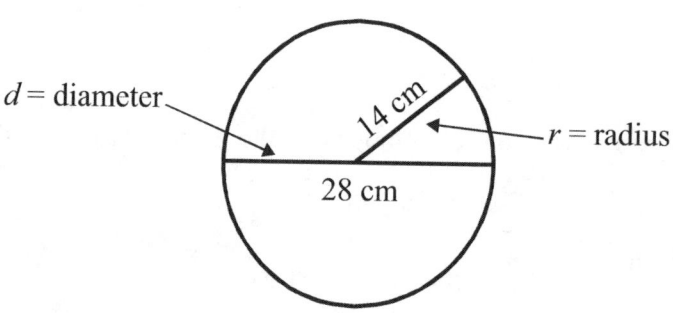

The formula for the circumference of a circle is $C = 2\pi r$ or $C = \pi d$. (The formulas are equal because the diameter is equal to twice the radius, $d = 2r$.)

Example 7:   Find the circumference of the circle above.

$C = \pi d$ $\quad\quad C = 2\pi r$
$C = \pi \times 28$ $\quad\quad C = 2 \times \pi \times 14$
$C = 28\pi$ cm $\quad\quad C = 28\pi$ cm

Use the formulas given above to find the circumferences of the following circles.

1. 8 in   C = _____
2. 14 ft   C = _____
3. 2 cm   C = _____
4. 6 m   C = _____
5. 8 ft   C = _____

6. 3 ft   C = _____
7. 12 in   C = _____
8. 6 m   C = _____
9. 5 cm   C = _____
10. 16 in   C = _____

12.9   Area of a Circle

The formula for the area of a circle is $A = \pi r^2$. The area is how many square units of measure would fit inside a circle.

Example 8:   Find the area of the circle, using both values for π.

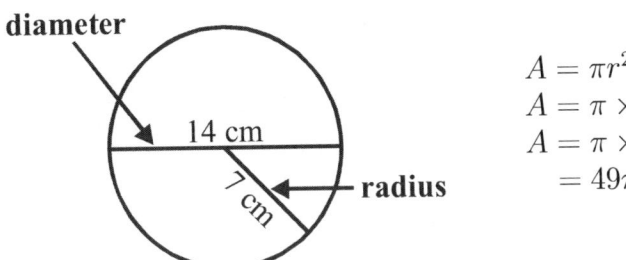

$A = \pi r^2$
$A = \pi \times 7^2$
$A = \pi \times 49$
$ = 49\pi \text{ cm}^2$

Find the area of the following circles. Remember to include units.

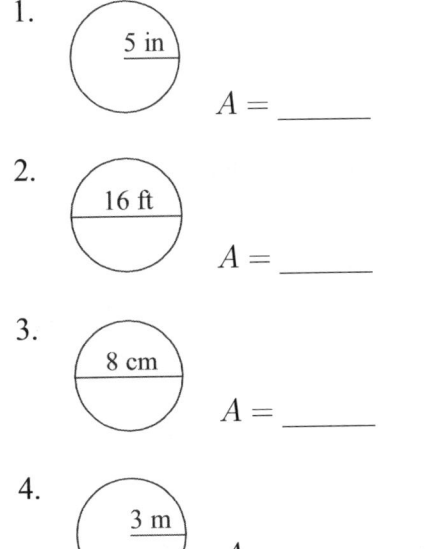

1. 5 in $A =$ _____
2. 16 ft $A =$ _____
3. 8 cm $A =$ _____
4. 3 m $A =$ _____

Fill in the chart below. Include appropriate units.

	Radius	Diameter	Area
5.	9 ft		
6.		4 in	
7.	8 cm		
8.		20 ft	
9.	14 m		
10.		18 cm	
11.	12 ft		
12.		6 in	

12.10 Two-Step Area Problems

Solving the problems below will require two steps. You will need to find the area of two figures, and then either add or subtract the two areas to find the answer. **Carefully read the examples.**

Example 9:
Find the area of the living room below.
Figure 1

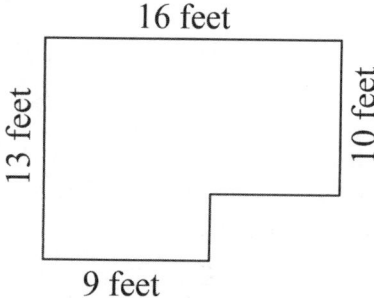

Step 1: Complete the rectangle as in Figure 2, and compute the area as if it were a complete rectangle.

Figure 2

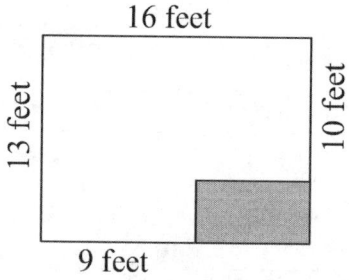

$A = \text{length} \times \text{width}$
$A = 16 \times 13$
$A = 208 \text{ ft}^2$

Step 2: Figure the area of the shaded part.

7 feet
3 feet
$7 \times 3 = 21 \text{ ft}^2$

Step 3: Subtract the area of the shaded part from the area of the complete rectangle.

$208 - 21 = 187 \text{ ft}^2$

Example 10:
Find the area of the shaded sidewalk.

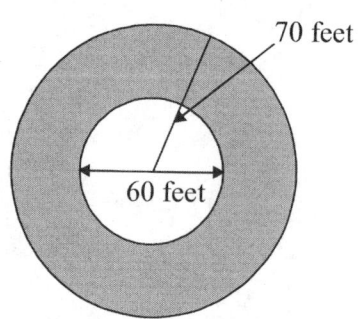

Step 1: Find the area of the outside circle.

$A = \pi \times 70 \times 70$
$A = 4900\pi \text{ ft}^2$

Step 2: Find the area of the inside circle.

$A = \pi \times 30 \times 30$
$A = 900\pi \text{ ft}^2$

Step 3: Subtract the area of the inside circle from the area of the outside circle.

$4900\pi - 900\pi = 4000\pi \text{ ft}^2$

12.10 Two-Step Area Problems

Find the area of the following figures.

1.
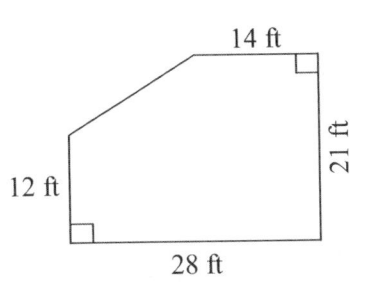

2.
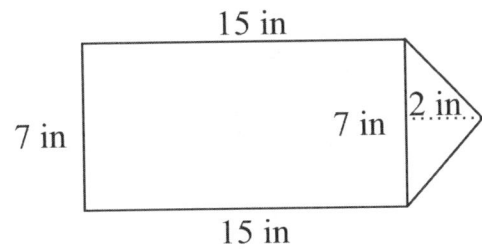

3. What is the area of the shaded circle?
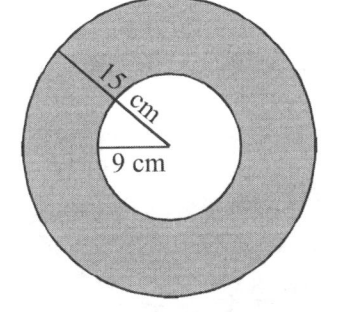

4.
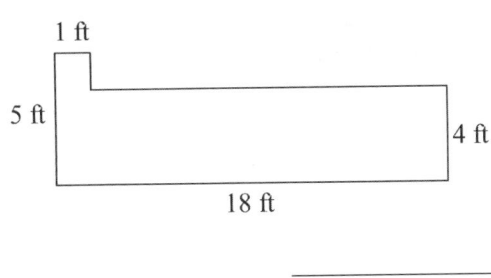

5.
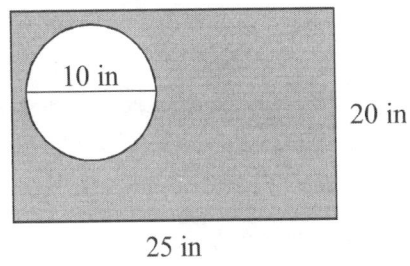

6. What is the area of the shaded part?
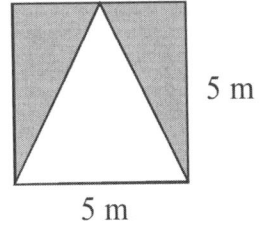

7. What is the area of the shaded part?
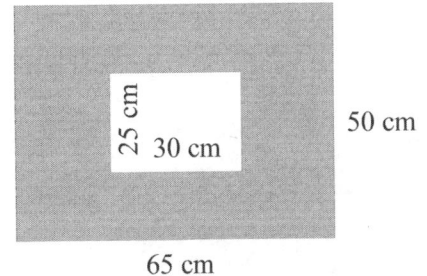

8.
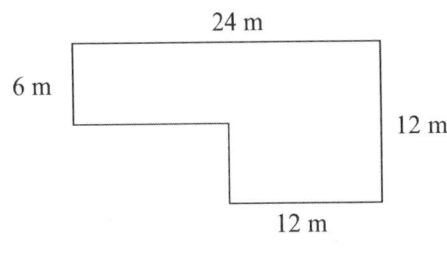

Chapter 12   Plane Geometry

12.11   Geometric Relationships of Plane Figures

This section illustrates what happens to the area of a figure when one or more of the dimensions is doubled or tripled.

Example 11: Sam drew a square that was 2 inches on each side for art class. His teacher said the square needed to be twice as big. When Sam doubled each side to 4 inches, what happened to the area?

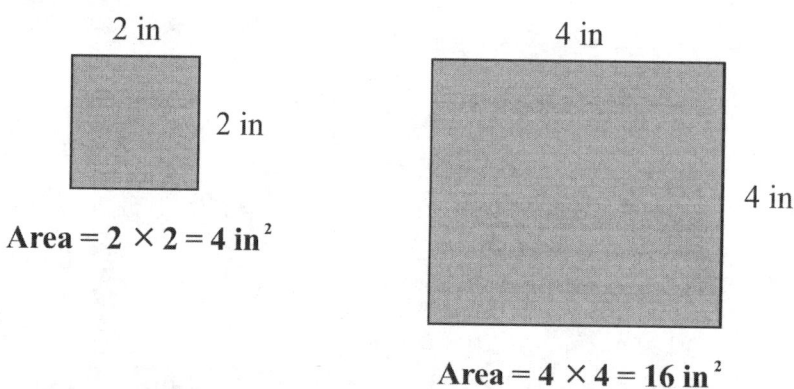

The area of the second square is 4 times larger than the first.

Example 12: Sonya drew a circle which had a radius of 3 inches for a school project. She also needed to make a larger circle which had a radius of 9 inches. When Sonya drew the bigger circle, what was the difference in area?

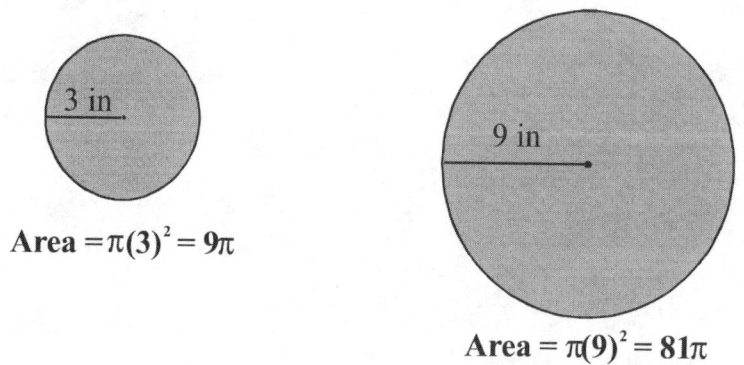

The area of the second circle is 9 times larger than the first.

From these two examples, we can determine that for every doubling or tripling of both sides or of the radius of a planar object, the total area increases by a squared value. In other words, when both sides of the square doubled, the area was 2^2 or 4 times larger. When the radius of the circle became 3 times larger, the area became 3^2 or 9 times larger.

12.11 Geometric Relationships of Plane Figures

Carefully read each problem below and solve.

1. Ken draws a circle with a radius of 5 cm. He then draws a circle with a radius of 10 cm. How many times larger is the area of the second circle?

2. Kobe draws a square with each side measuring 6 inches. He then draws a rectangle with a width of 6 inches and a length of 12 inches. How many times larger is the area of the rectangle than the area of the square? (**Hint:** The increase is *not* equal in both directions.)

3. Toshi draws a square 3 inches on each side. Then he draws a bigger square that is 6 inches on each side. How many times larger is the area of the second square than the area of the first square?

4. Leslie draws a triangle with a base of 5 inches and a height of 3 inches. To use her triangle pattern for a bulletin board design, it needs to be 3 times bigger. If she increases the base and the height by multiplying each by 3, how much will the area of the triangle increase?

5. Heather is using 100 tiles that measure 1 foot by 1 foot to cover a 10 feet by 10 feet floor. If she used tiles that measure 2 feet by 2 feet, how many tiles would she need?

6. The area of circle B is 9 times larger than the area of circle A. If the radius of circle A is represented by x, how would you represent the radius of circle B?

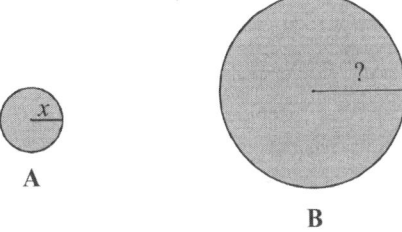

7. How many squares will it take to fill the rectangle below?

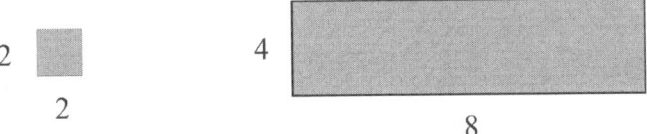

8. If the area of diamond B is one-fourth the area of diamond A, what are the dimensions of diamond B?

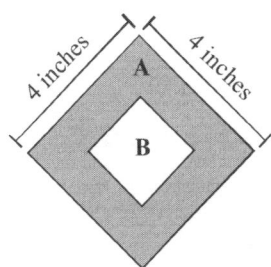

Chapter 12 Review

1. Find the area of the shaded region of the figure below.

   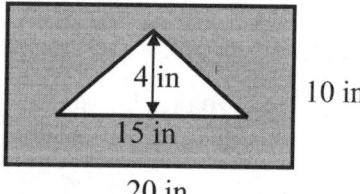

2. Calculate the perimeter.

   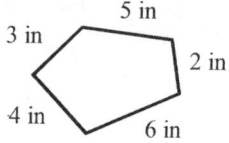

3. Calculate the perimeter and area of the following figure.

   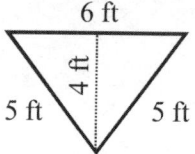

4. Find the area of the shaded part.

   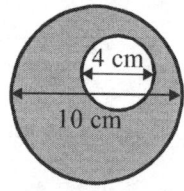

5. Calculate the circumference and the area of the following circle.

   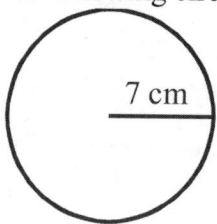

6. If you double the side of a square, how much does the area of the square increase?

7. What is the area of a square which measures 8 inches on each side?

8. Calculate the circumference and the area of the following circle.

   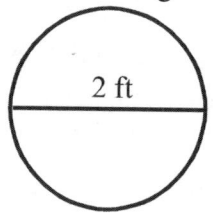

9. What is the sum of the measures of the interior angles in the figure below?

   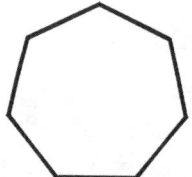

10. Find the area of the trapezoid below.

    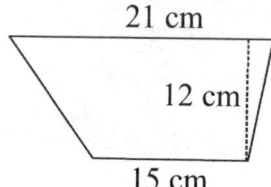

11. Find the area of the parallelogram below.

    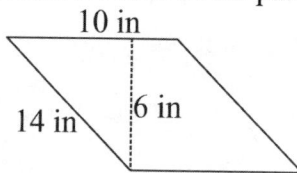

12. Find the area of the rhombus below.

    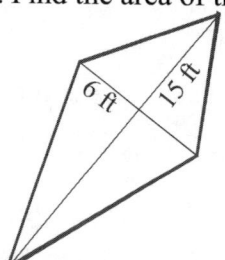

Chapter 12 Test

1. What is the area of a circle with a radius of 7 cm?

   (A) 49π square cm
   (B) 62π square cm
   (C) 14π square cm
   (D) 196π square cm

2. How many square feet of sod are needed to cover a 9-foot by 60-foot lawn?

   (A) 69 square feet
   (B) 138 square feet
   (C) 270 square feet
   (D) 540 square feet

3. The figure below is a circle inscribed in a square. What is the area of the shaded region?

   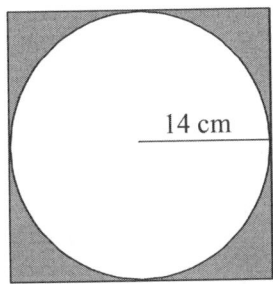

   (A) 784 square centimeters
   (B) $784 - 196\pi$ square centimeters
   (C) $784 - 28\pi$ square centimeters
   (D) $784 - 14\pi$ square centimeters

4. Using the formula $A = \frac{1}{2}bh$ for the area of a triangle, find the area of the triangle below.

   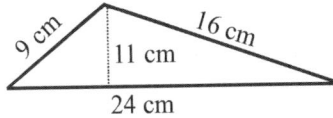

   (A) 49 cm^2
   (B) 132 cm^2
   (C) 264 cm^2
   (D) 3,456 cm^2

5. What is the sum of the measures of the interior angles of a pentagon?

   (A) 180°
   (B) 360°
   (C) 540°
   (D) 900°

6. Which of the following figures is a parallelogram?

   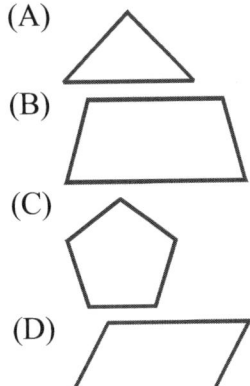

7. What is the area of the figure below?

   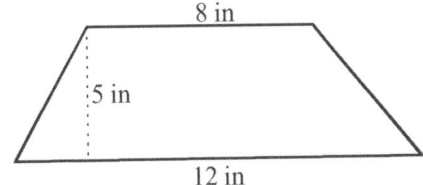

   (A) 25 in^2
   (B) 50 in^2
   (C) 100 in^2
   (D) 480 in^2

8. What is the perimeter of a hexagon if each side measures 8 inches?

   (A) 48 inches
   (B) 40 inches
   (C) 32 inches
   (D) 24 inches

Copyright © American Book Company

Chapter 13
Solid Geometry

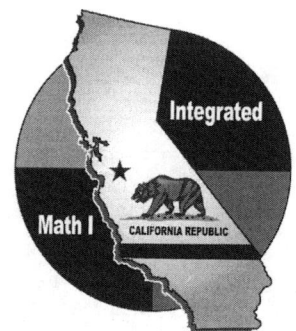

This chapter covers the following CA Integrated Math I standards:

Geometry	9.0
	8.0
	11.0

13.1 Understanding Volume

Volume - Measurement of volume is expressed in cubic units such as in^3, ft^3, m^3, cm^3, or mm^3. The volume of a solid is the number of cubic units that can be contained in the solid.

First, let's look at rectangular solids.

Example 1: How many 1 cubic centimeter cubes will it take to fill up the figure below?

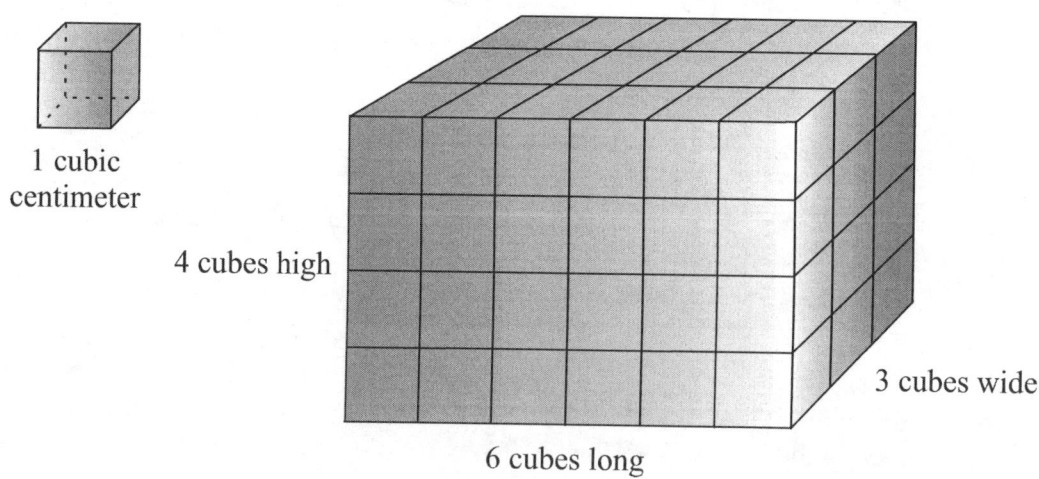

To find the volume, you need to multiply the length times the width times the height.
Volume of a rectangular solid = length × width × height ($V = lwh$).
$$V = 6 \times 3 \times 4 = 72 \text{ cm}^3$$

13.2 Volume of Rectangular Prisms and Cubes

You can calculate the volume (V) of a rectangular prism (box) by multiplying the length (l) by the width (w) by the height (h), as expressed in the formula $V = (lwh)$. A **cube** is a special kind of rectangular prism (box). Each side of a cube has the same measure. So, the formula for the volume of a cube is $V = s^3$ ($s \times s \times s$).

Example 2: Find the volume of the box pictured here:

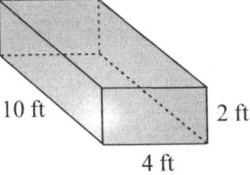

Step 1: Insert measurements from the figure into the formula.
Step 2: Multiply to solve. $10 \times 4 \times 2 = 80$ ft^3

Note: Volume is always expressed in cubic units such as in^3, ft^3, m^3, cm^3, or mm^3.

Find the volume of the following rectangular prisms (boxes).

1.
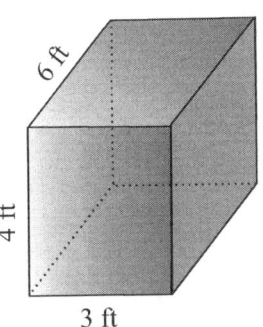

2.
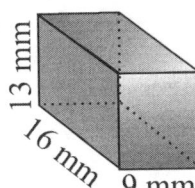

3.
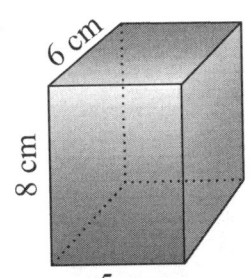

4.
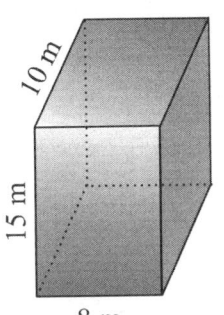

5.
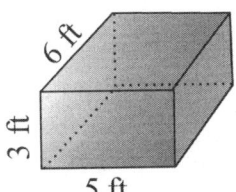

6.
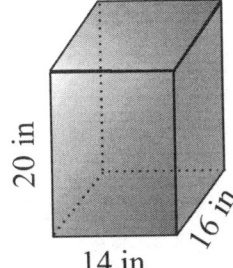

7.
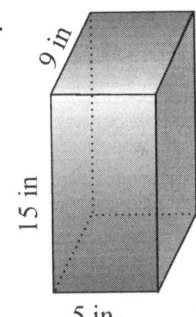

8.
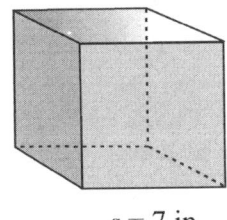
$s = 7$ in.

9.
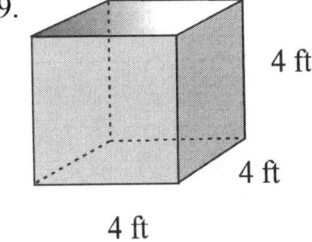

Chapter 13   Solid Geometry

13.3   Volume of Spheres, Cones, Cylinders, and Pyramids

To find the volume of a solid, insert the measurements given for the solid into the correct formula and solve. Remember, volumes are expressed in cubic units such as in^3, ft^3, m^3, cm^3, or mm^3.

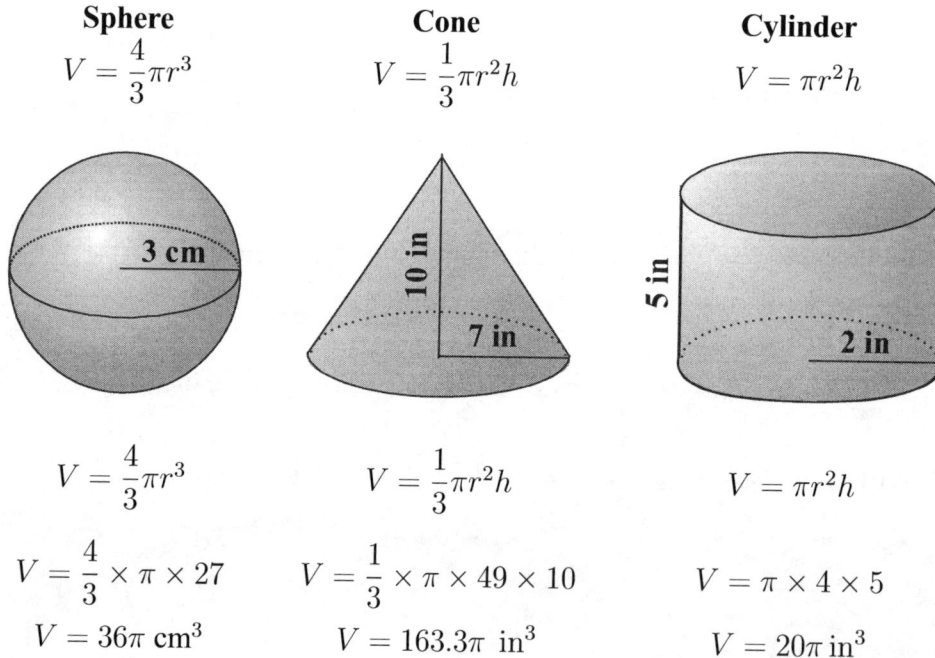

Sphere	Cone	Cylinder
$V = \frac{4}{3}\pi r^3$	$V = \frac{1}{3}\pi r^2 h$	$V = \pi r^2 h$

$V = \frac{4}{3}\pi r^3 \qquad\qquad V = \frac{1}{3}\pi r^2 h \qquad\qquad V = \pi r^2 h$

$V = \frac{4}{3} \times \pi \times 27 \qquad V = \frac{1}{3} \times \pi \times 49 \times 10 \qquad V = \pi \times 4 \times 5$

$V = 36\pi \text{ cm}^3 \qquad\qquad V = 163.3\pi \text{ in}^3 \qquad\qquad V = 20\pi \text{ in}^3$

Pyramids

$V = \frac{1}{3}Bh$ B = area of rectangular base $\qquad\qquad V = \frac{1}{3}Bh$ B = area of triangular base

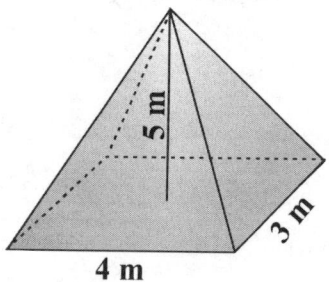 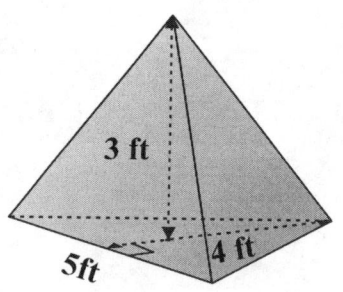

$V = \frac{1}{3}Bh \quad B = l \times w \qquad\qquad\qquad B = \frac{1}{2} \times 5 \times 4 = 10 \text{ ft}^2$

$V = \frac{1}{3} \times 4 \times 3 \times 5 \qquad\qquad\qquad V = \frac{1}{3} \times 10 \times 3$

$V = 20 \text{ m}^3 \qquad\qquad\qquad\qquad\qquad V = 10 \text{ ft}^3$

13.3 Volume of Spheres, Cones, Cylinders, and Pyramids

Find the volume of the following shapes.

1.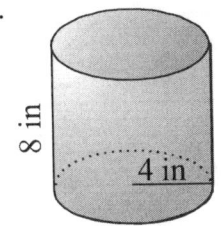

2.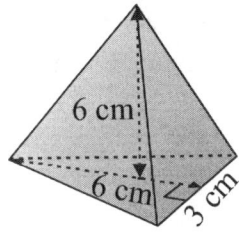

3.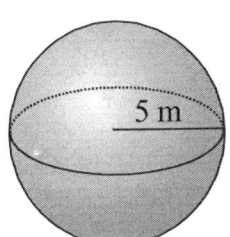

4.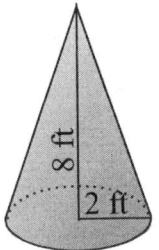

5.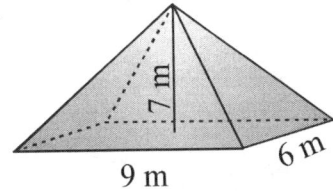

6.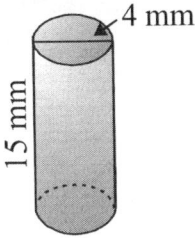

7.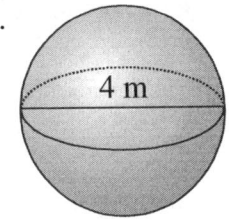

8.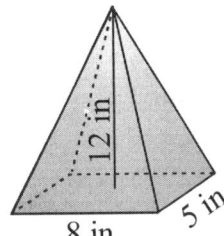

9.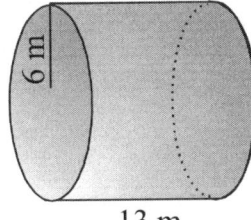

10.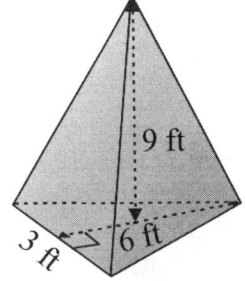

Chapter 13   Solid Geometry

13.4   Two-Step Volume Problems

Some objects are made from two geometric figures. For example, the tower below is made up of two geometric objects, a rectangular prism and a pyramid.

Example 3:   Find the maximum volume of the tower.

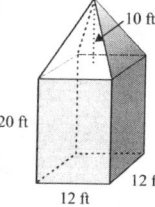

Step 1:   Determine which formulas you will need. The tower is made from a pyramid and a rectangular prism, so you will need the formulas for the volume of these two figures.

Step 2:   Find the volume of each part of the tower. The bottom of the tower is a rectangular prism $V = lwh$
$V = 12 \times 12 \times 20 = 2,880 \text{ ft}^3$
The top of the tower is a rectangular pyramid. $V = \frac{1}{3}Bh$
$V = \frac{1}{3} \times 12 \times 12 \times 10 = 480 \text{ ft}^3$

Step 3:   Add the two volumes together. $2800 \text{ ft}^3 + 480 \text{ ft}^3 = 3,360 \text{ ft}^3$

Find the volume of the geometric figures below. Hint: If part of a solid has been removed, find the volume of the hole, and subtract it from the volume of the total object.

1.

(figure: stepped solid with labels 8 in, 8 in, 8 in, 16 in, 8 in, 8 in)

2. Each side measures 3 inches.

3. A rectangular hole passes through the middle of the figure below. The hole measures 1 cm on each side.

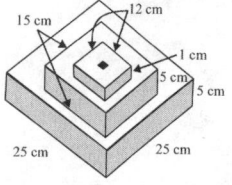

4. In the figure below, 3 cylinders are stacked on top of each other. The radii of the cylinders are 2 inches, 4 inches, and 6 inches. The height of each cylinder is 1 inch.

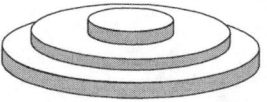

5.

(figure: stepped solid with labels 2 cm, 3 cm, 3 cm, 3 cm, 3 cm, 3 cm, 8 cm, 10 cm, 3 cm)

6. A hole, 1 meter in diameter, has been cut through the cylinder below.

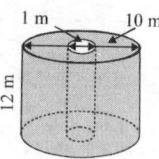

13.5 Geometric Relationships of Solids

In the previous chapter, you looked at geometric relationships between 2-dimensional figures. Now you will learn about the relationships between 3-dimensional figures. The formulas for finding the volumes of geometric solids are given below.

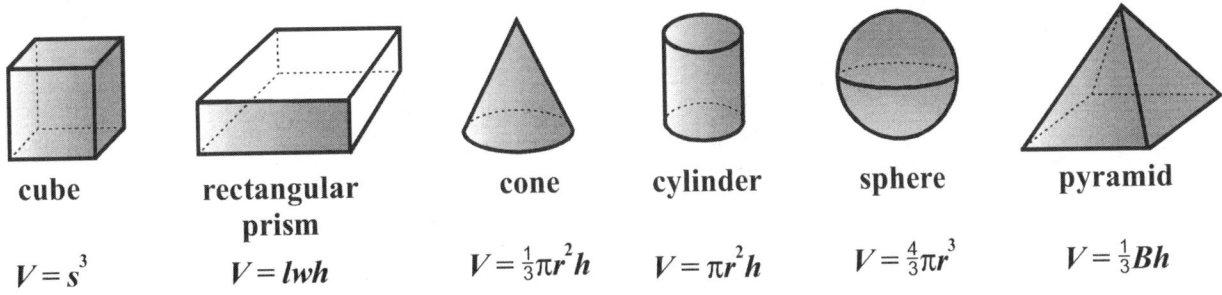

cube	rectangular prism	cone	cylinder	sphere	pyramid
$V = s^3$	$V = lwh$	$V = \frac{1}{3}\pi r^2 h$	$V = \pi r^2 h$	$V = \frac{4}{3}\pi r^3$	$V = \frac{1}{3}Bh$

By studying each formula and by comparing formulas between different solids, you can determine general relationships.

Example 4: How would doubling the radius of a sphere affect the volume?

The volume of a sphere is $v = \frac{4}{3}\pi r^3$. Just by looking at the formula, can you see that by doubling the radius, the volume would increase 8 times the original volume? So, a sphere with a radius of 2 would have a volume 8 times greater than a sphere with a radius of 1.

Example 5: A cylinder and a cone have the same radius and the same height. What is the difference between their volumes?

Compare the formulas for the volume of a cone and the volume of a cylinder. They are identical except that the cone is multiplied by $\frac{1}{3}$. Therefore, the volume of a cone with the same height and radius as a cylinder would be one-third less. Or, the volume of a cylinder with the same height and radius as a cone would be three times greater.

Example 6: If you double one dimension of a rectangular prism, how will the volume be affected? How about doubling two dimensions? How about doubling all three dimensions?

Do you see that doubling just one of the dimensions of a rectangular prism will also double the volume? Doubling two of the dimensions will cause the volume to increase 4 times the original volume. Doubling all three dimensions will cause the volume to increase 8 times the original volume.

Example 7: A cylinder holds 100 cubic centimeters of water. If you triple the radius of the cylinder but keep the height the same, how much water would you need to fill the new cylinder?

Tripling the radius of a cylinder causes the volume to increase by 3^2 or 9 times the original volume. The volume of the new cylinder would hold 9×100 or 900 cubic centimeters of water.

Chapter 13   Solid Geometry

Answer the following questions by comparing the volumes of two solids that share some of the same dimensions.

1. If you have a cylinder with a height of 8 inches and a radius of 4 inches, and you have a cone with the same height and radius, how many times greater is the volume of the cylinder than the volume of the cone?

2.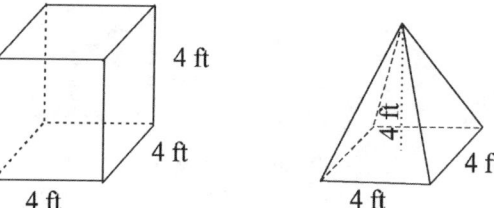

    In the two figures above, how many times larger is the volume of the cube than the volume of the pyramid?

3. How many times greater is the volume of a cylinder if you double the radius?

4. How many times greater is the volume of a cylinder if you double the height?

5. In a rectangular solid, how many times greater is the volume if you double the length?

6. In a rectangular solid, how many times greater is the volume if you double the length and the width?

7. In a rectangular solid, how many times greater is the volume if you double the length and the width and the height?

8. In the following two figures, how many cubes like Figure 1 will fit inside Figure 2?

    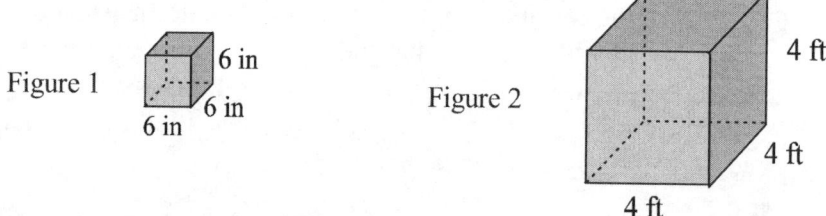

9. A sphere has a radius of 1. If the radius is increased to 3, how many times greater will the volume be?

10. It takes 2 liters of water to fill cone A below. If the cone is stretched so the radius is doubled, but the height stays the same, how much water is needed to fill the new cone, B?

    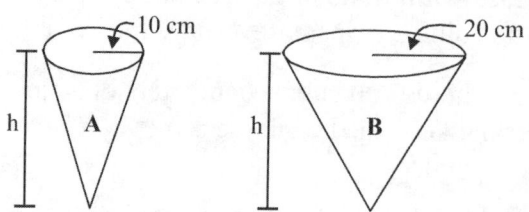

13.6 Surface Area

The **surface area of a solid** is the total area of all the sides of a solid.

13.7 Cube

There are six sides on a cube. To find the surface area of a cube, find the area of one side and multiply by 6.

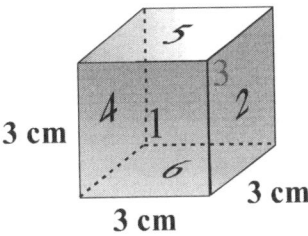

Area of each side of the cube: $3 \times 3 = 9 \text{ cm}^2$
Total surface area: $9 \times 6 = 54 \text{ cm}^2$

13.8 Rectangular Prisms

There are 6 sides on a rectangular prism. To find the surface area, add the areas of the six rectangular sides.

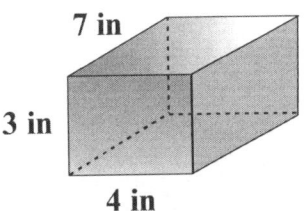

Top and Bottom

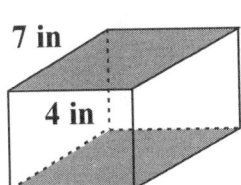

Area of top side:
$7 \text{ in} \times 4 \text{ in} = 28 \text{ in}^2$
Area of top and bottom:
$28 \text{ in} \times 2 = 56 \text{ in}^2$

Front and Back

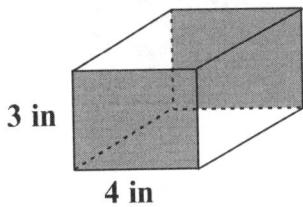

Area of front:
$3 \text{ in} \times 4 \text{ in} = 12 \text{ in}^2$
Area of front and back:
$12 \text{ in} \times 2 = 24 \text{ in}^2$

Left and Right

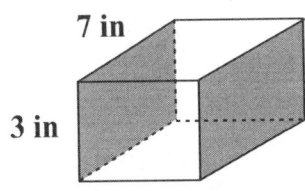

Area of left side:
$3 \text{ in} \times 7 \text{ in} = 21 \text{ in}^2$
Area of left and right:
$21 \text{ in} \times 2 = 42 \text{ in}^2$

Total surface area: $56 \text{ in}^2 + 24 \text{ in}^2 + 42 \text{ in}^2 = 122 \text{ in}^2$

Chapter 13   Solid Geometry

Find the surface area of the following cubes and prisms.

1.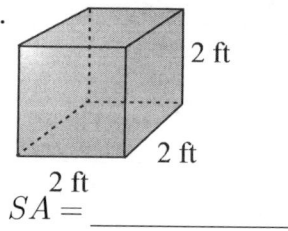
SA = _____

6.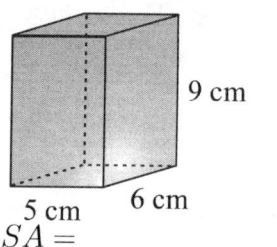
SA = _____

2.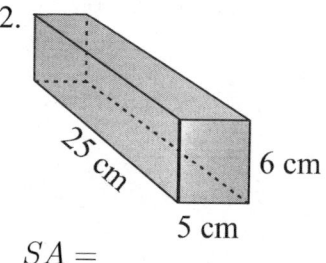
SA = _____

7.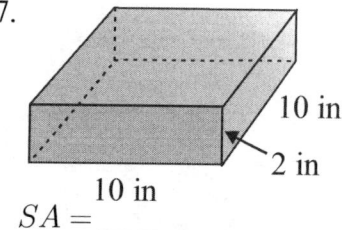
SA = _____

3.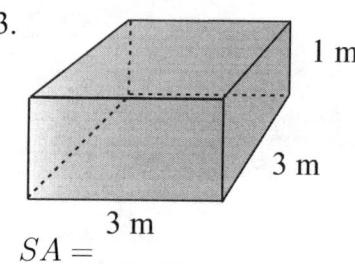
SA = _____

8.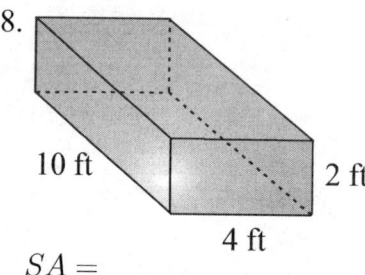
SA = _____

4.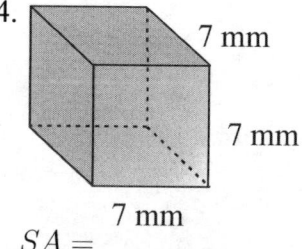
SA = _____

9.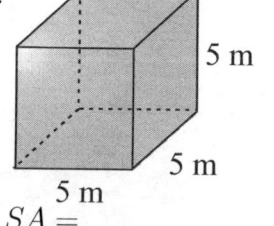
SA = _____

5.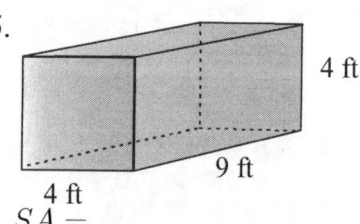
SA = _____

10.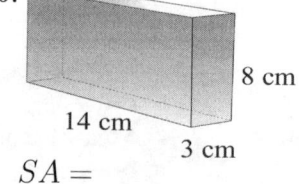
SA = _____

13.9 Pyramid

The pyramid below is made of a square base with 4 triangles on the sides.

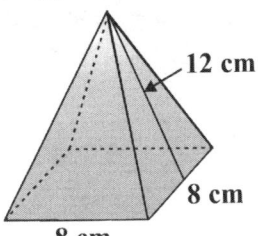
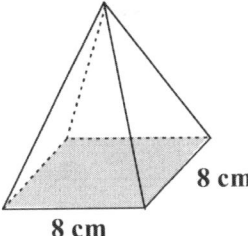
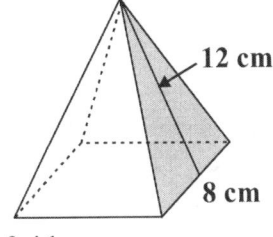

Area of square base:
$A = l \times w$
$A = 8 \times 8 = 64 \text{ cm}^2$

Area of sides:
Area of 1 side $= \frac{1}{2}bh$
$A = \frac{1}{2} \times 8 \times 12 = 48 \text{ cm}^2$
Area of 4 sides $= 48 \times 4 = 192 \text{ cm}^2$

Total surface area: $64 + 192 = 256 \text{ cm}^2$

Find the total surface of the following pyramids.

1.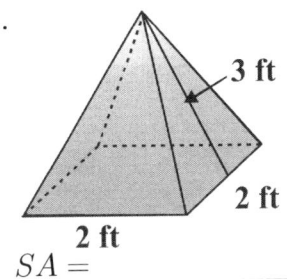
SA = _____

2.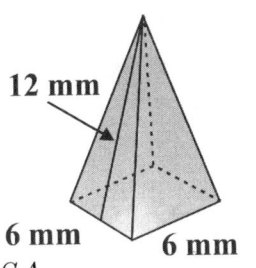
SA = _____

3.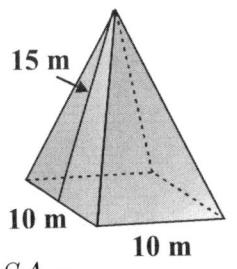
SA = _____

4.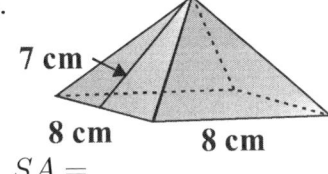
SA = _____

5.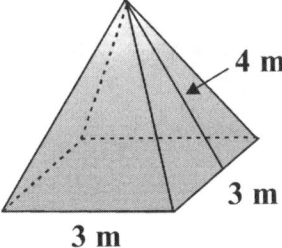
SA = _____

6.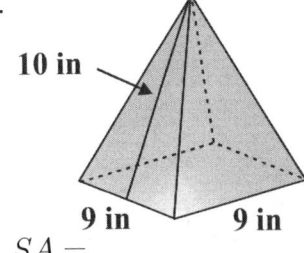
SA = _____

7.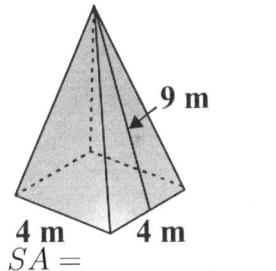
SA = _____

8.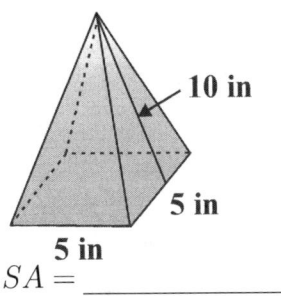
SA = _____

9.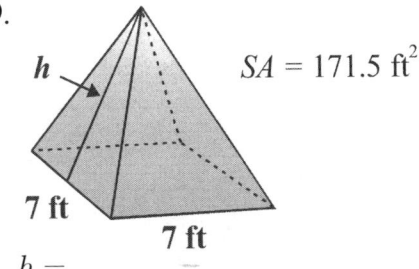
SA = 171.5 ft^2
h = _____

13.10 Cylinder

If the side of a cylinder was slit from top to bottom and laid flat, its shape would be a rectangle. The length of the rectangle is the same as the circumference of the circle that is the base of the cylinder. The width of the rectangle is the height of the cylinder.

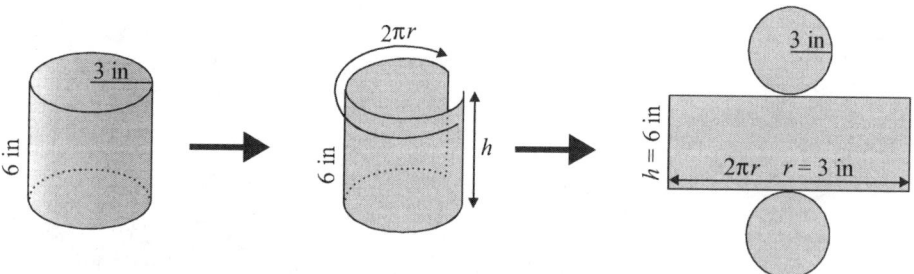

Total Surface Area of a Cylinder $= 2\pi r^2 + 2\pi rh$

Area of top and bottom:
Area of a circle $= \pi r^2$
Area of top $= \pi \times 3^2 = 9\pi$ in^2
Area of top and bottom $= 2 \times 9\pi = 18\pi$ in^2

Area of side:
Area of rectangle $= l \times h$
$l = 2\pi r = 2 \times \pi \times 3 = 6\pi$ in
Area of rectangle $= 6\pi \times 6 = 36\pi$ in^2

Total surface area $= 18\pi + 36\pi = 54\pi$ in^2

Find the total surface area of the following cylinders.

1.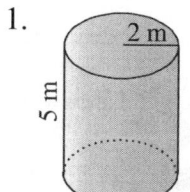

2.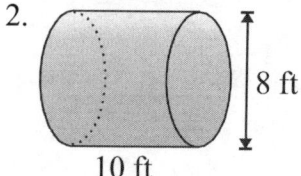

3.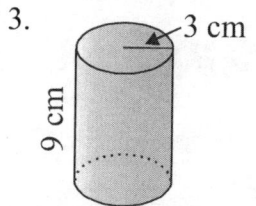

4.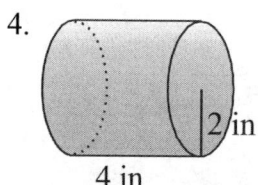

5.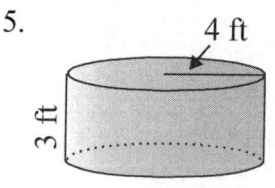

6.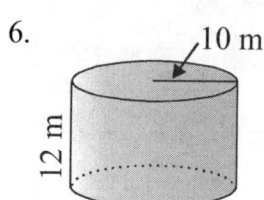

7.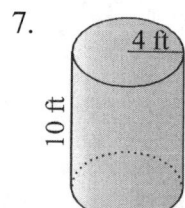

8.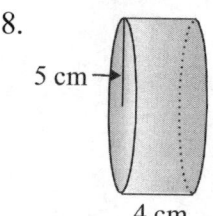

9.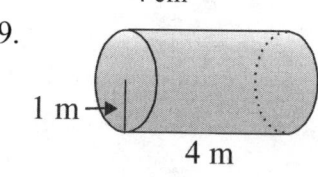

13.11 Sphere

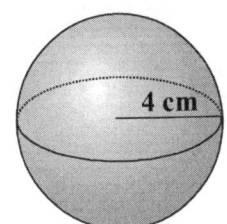

Surface area $= 4\pi r^2$
Surface area $= 4 \times \pi \times 4^2$
Surface area $= 64\pi$ cm^2

Find the surface area of a sphere given the following measurements where $r =$ radius and $d =$ diameter.

1. $r = 2$ in $SA =$ _____
2. $r = 6$ m $SA =$ _____
3. $r = \frac{3}{4}$ yd $SA =$ _____
4. $d = 8$ cm $SA =$ _____
5. $d = 50$ mm $SA =$ _____
6. $r = \frac{1}{4}$ ft $SA =$ _____
7. $d = 14$ cm $SA =$ _____
8. $r = \frac{1}{5}$ km $SA =$ _____
9. $d = 3$ in $SA =$ _____
10. $d = \frac{2}{3}$ ft $SA =$ _____
11. $r = 10$ mm $SA =$ _____
12. $d = 5$ yd $SA =$ _____

13.12 Cone

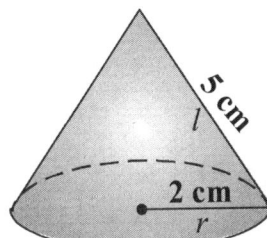

Total Surface Area: $T = \pi r (r + s)$
$r =$ radius of base $s =$ slant height
$T = \pi \times 2(2 + 5)$
$T = 2\pi \times 7$
$T = 14\pi$ cm^2

Find the surface area of the following cones.

1.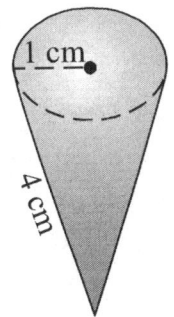

2.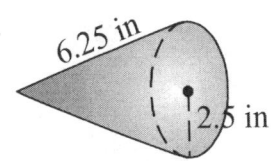

3.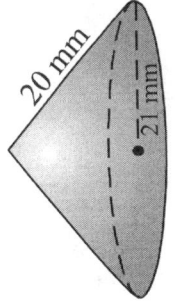

4.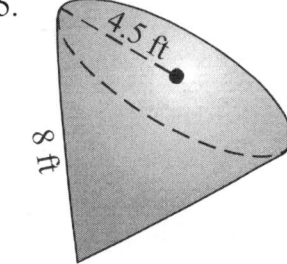

5.

6.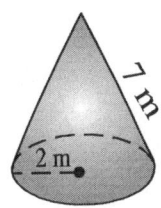

Chapter 13　Solid Geometry

13.13　Solid Geometry Word Problems

1. If an Egyptian pyramid has a square base that measures 500 yards by 500 yards, and the pyramid stands 300 yards tall, what would be the volume of the pyramid? Use the formula for volume of a pyramid, $V = \frac{1}{3}Bh$ where B is the area of the base.

2. Robert is using a cylindrical barrel filled with water to flatten the sod in his yard. The circular ends have a radius of 1 foot. The barrel is 3 feet wide. How much water will the barrel hold? The formula for volume of a cylinder is $V = \pi r^2 h$.

3. If a basketball measures 24 centimeters in diameter, what volume of air will it hold? The formula for volume of a sphere is $V = \frac{4}{3}\pi r^3$.

4. What is the volume of a cone that is 2 inches in diameter and 5 inches tall? The formula for volume of a cone is $V = \frac{1}{3}\pi r^2 h$.

5. Kelly has a rectangular fish aquarium that measures 24 inches wide, 12 inches deep, and 18 inches tall. What is the maximum amount of water that the aquarium will hold?

6. Jenny has a rectangular box that she wants to cover in decorative contact paper. The box is 10 cm long, 5 cm wide, and 5 cm high. How much paper will she need to cover all 6 sides?

7. Gasco needs to construct a cylindrical, steel gas tank that measures 6 feet in diameter and is 8 feet long. How many square feet of steel will be needed to construct the tank? Use the following formulas as needed: $A = l \times w$, $A = \pi r^2$, $C = 2\pi r$.

8. Craig wants to build a miniature replica of San Francisco's Transamerica Pyramid out of glass. His replica will have a square base that measures 6 cm by 6 cm. The 4 triangular sides will be 6 cm wide and 60 cm tall. How many square centimeters of glass will he need to build his replica? Use the following formulas as needed: $A = l \times w$ and $A = \frac{1}{2}bh$.

9. Jeff built a wooden, cubic toy box for his son. Each side of the box measures 2 feet. How many square feet of wood did he use to build the toy box? How many cubic feet of toys will the box hold?

Chapter 13 Review

Find the volume and/or surface area of the following solids.

1.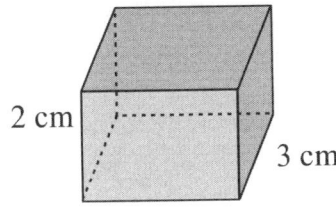
V = _____
SA = _____

2.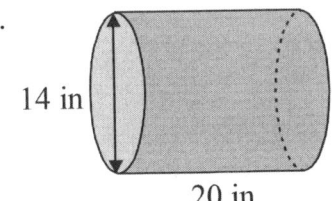
V = _____
SA = _____

3.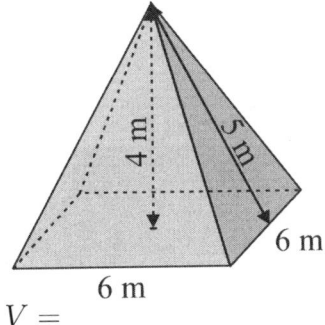
V = _____
SA = _____

4.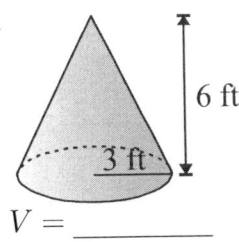
V = _____

5.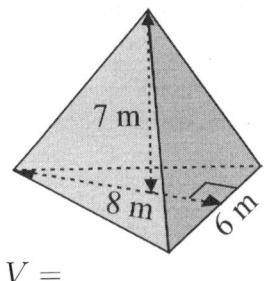
V = _____

6.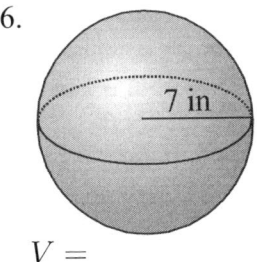
V = _____
SA = _____

7. The sandbox at the local elementary school is 60 inches wide and 100 inches long. The sand in the box is 6 inches deep. How many cubic inches of sand are in the sandbox?

8. If you have cubes that are two inches on each edge, how many would fit in a cube that was 16 inches on each edge?

9. If you double each edge of a cube, how many times larger is the volume?

10.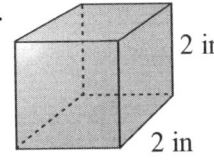
It takes 8 cubic inches of water to fill the cube above. If each side of the cube is doubled, how much water is needed to fill the new cube?

11. If a ball is 4 inches in diameter, what is its surface area?

12. A grain silo is in the shape of a cylinder. If the silo has an inside diameter of 10 feet and a height of 35 feet, what is the maximum volume inside the silo?

13. A closed cardboard box is 30 centimeters long, 10 centimeters wide, and 20 centimeters high. What is the total surface area of the box?

14. Siena wants to build a wooden toy box with a lid. The dimensions of the toy box are 3 feet long, 4 feet wide, and 2 feet tall. How many square feet of wood will she need to construct the box?

15. How many 1-inch cubes will fit inside a larger 1 foot cube? (Figures are not drawn to scale.)

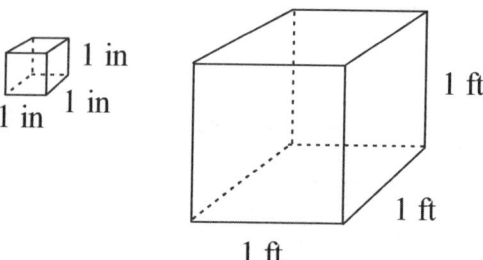

16. Estimate the volume of the figure below.

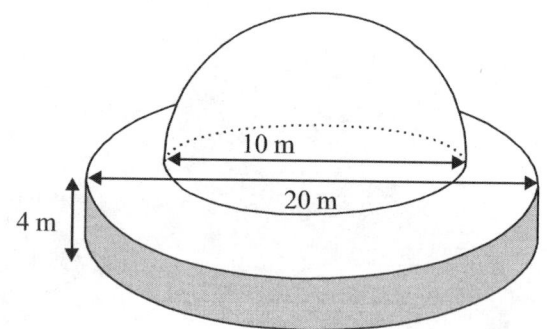

Chapter 13 Test

1. What is the volume, in cubic feet, of the square pyramid below?

   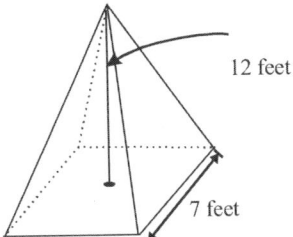

   (A) 168 cubic feet
   (B) 196 cubic feet
   (C) 294 cubic feet
   (D) 588 cubic feet

2. What is the volume of the following oil tank? Round your answer to the nearest hundredth.

   Use the formula $V = \pi r^2 h$

   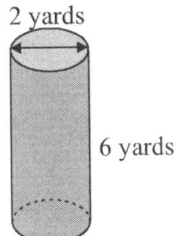

   (A) 6π yd^3
   (B) 12π yd^3
   (C) 24π yd^3
   (D) 48π yd^3

3. Jack is going to paint the ceiling and four walls of a room that is 10 feet wide, 12 feet long, and 10 feet from floor to ceiling. How many square feet will he paint?

   (A) 120 square feet
   (B) 560 square feet
   (C) 68 square feet
   (D) 1,200 square feet

4. What is the volume of a wading pool 12 feet long, 6 feet wide, and 6 inches deep?

   (A) 18 cubic feet
   (B) 36 cubic feet
   (C) 216 cubic feet
   (D) 432 cubic feet

5. What is the volume of the figure below?

   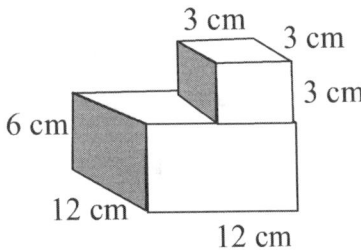

   (A) 21 cm
   (B) 39 cm
   (C) 216 cm
   (D) 891 cm

6. Use the formula for volume of a sphere to determine the volume of the hemisphere below. Volume of a sphere $= \frac{4}{3}\pi r^3$.

   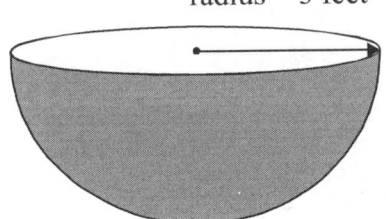

   (A) 4π
   (B) 12π
   (C) 18π
   (D) 36π

7. Find the volume of the cone. Use the formula $V = \frac{1}{3}\pi r^2 h$.

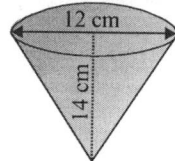

(A) 1512π cm³
(B) 84π cm³
(C) 168π cm³
(D) 504π cm³

8. What is the volume of the box shown below?

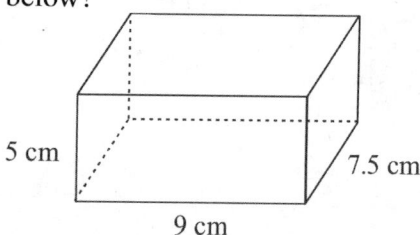

(A) 22.5 cm³
(B) 337.5 cm³
(C) 1875 cm³
(D) 2250 cm³

9. Compute the entire surface area of the three-dimensional object shown below.

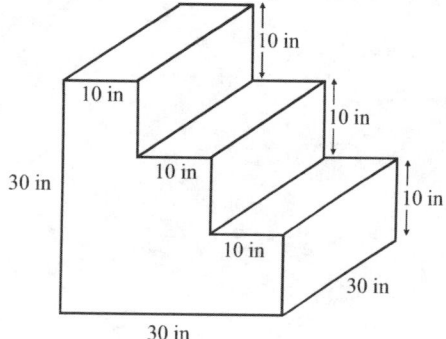

(A) 4800 in²
(B) 5000 in²
(C) 5200 in²
(D) 5400 in²

10. The ratio of two similar spheres' radii is 1 to 3. What is the ratio of their volumes?

(A) $\frac{2}{3}$
(B) $\frac{1}{3}$
(C) $\frac{1}{9}$
(D) $\frac{1}{27}$

11. Daniel of Daniel's Ice Cream is trying to figure out how big to order the wrappers that go around the waffle cones he sells. In order to figure this out, he must find the surface area of the waffle cone shown below. The cone is three inches in diameter. What will it's lateral area approximately be?

(A) 7.5 in²
(B) 15π in²
(C) 7.5π in²
(D) 9.75π in²

12. Use the formula for surface area of a sphere and area of a circle to determine the surface area of the hemisphere below. Surface area of a sphere $= 4\pi r^2$.

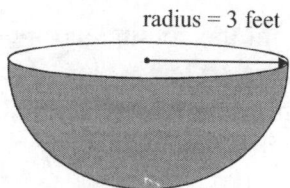

(A) 18π
(B) 27π
(C) 24π
(D) 45π

Practice Test 1

1. For this question, use the following diagram of a cube with an edge length of 2.

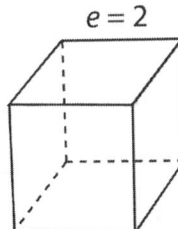

Which of the following shapes have the same volume?

(A)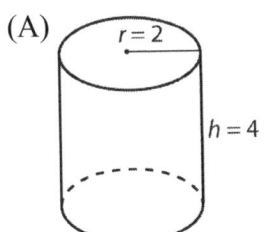

(B)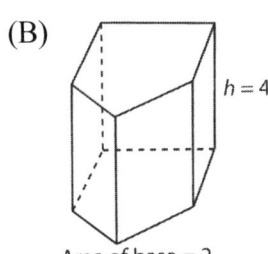

(C)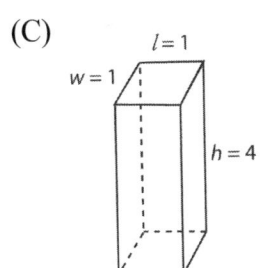

(D)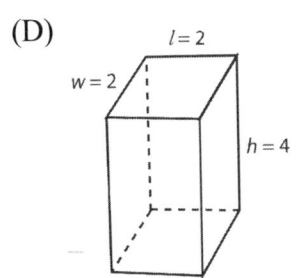

G9.0

2. Consider the following equations:

$f(x) = 3x + 2$ and $f(x) = 3x - 7$.

Which of the following statements is true concerning the graphs of these equations?

(A) The graphs of the equations are lines that are perpendicular to each other.

(B) The graph of the line represented by the equation $f(x) = 3x + 2$ always remains above the x-axis, while the graph of the line represented by the equation $f(x) = 3x - 7$ always remains below the x-axis.

(C) The graphs of the equations are lines that are parallel to each other, but that have different y-intercepts.

(D) The graphs of the lines intersect each other at the point $(2, -7)$.

A8.0

3. What is the distance between points $(-2, -4)$ and $(4, -7)$?

(A) $\sqrt{3}$
(B) 45
(C) 3
(D) $3\sqrt{5}$

G17.0

4. Solve by factoring:

$11x^2 - 31x - 6 = 0$

(A) 3 and -2
(B) 3 and $-\frac{2}{11}$
(C) 3 and $-\frac{11}{2}$
(D) -3 and $\frac{11}{2}$

A14.0

Copyright © American Book Company

5. The figure below is an isosceles triangle with two parallel lines within it. Find the measure of angle FGC.

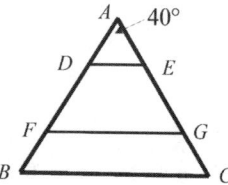

(A) 70°
(B) 140°
(C) 110°
(D) 290°

6. Add: $\dfrac{2y}{x} + \dfrac{2x-y}{y}$

(A) $\dfrac{2x^2 - xy + 2y^2}{xy}$

(B) $\dfrac{2x+y}{xy}$

(C) $\dfrac{4xy + 2y^2}{xy}$

(D) $\dfrac{2x^2 - xy + 2y^2}{x+y}$

7. Solve: $\dfrac{3x+6}{-2} > -12$

(A) $x < 24$
(B) $x > 0$
(C) $x > 6$
(D) $x < 6$

8. Solve for x: $2(x+5) + 4(2x-1) = -14$

(A) $x = -2$
(B) $x = -1$
(C) $x = -1\frac{4}{5}$
(D) $x = -1\frac{2}{10}$

9. Simplify: $\dfrac{(3a^2)^3}{a^3}$

(A) $27a^3$
(B) $\dfrac{9a^6}{a^3}$
(C) $9a^3$
(D) $\dfrac{3a^6}{a^3}$

10. Solve by using the quadratic formula: $y^2 - 4y - 12 = 0$

(A) $\{2, -6\}$
(B) $\{-2, 6\}$
(C) $\{3, -4\}$
(D) $\{-3, 4\}$

11. Someone reports a fire at the location plotted on the grid.

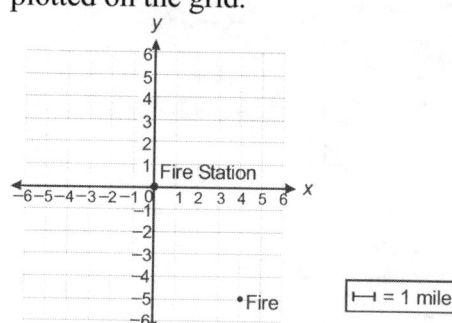

How far is the fire from the fire station?

(A) 3 miles
(B) $\sqrt{20}$ miles
(C) $\sqrt{41}$ miles
(D) 9 miles

12. Solve for a: $-4a - 12 = -36$

(A) 6
(B) -6
(C) 12
(D) -12

13. Find $(4y^4 + 2y^2 + 7) + (2y^3 + 5y^2 - 4)$.

   (A) $4y^4 + 2y^3 + 7y^2 + 3$
   (B) $4y^4 + 4y^3 + 5y^2 + 3$
   (C) $8y^7 + 10y^4 - 28$
   (D) $8y^{12} + 10y^4 + 3$

   A10.0

14. Find $(-3a^2 + 8a - 2) = (-4a^2 - 2a + 6)$.

   (A) $a^2 + 10a - 8 = 0$
   (B) $-7a^2 + 6a + 4 = 0$
   (C) $12a^4 - 16a^2 - 12 = 0$
   (D) $a^2 + 6a - 8 = 0$

   A10.0

15. Use the correct order of operations to evaluate the following expression.
    $4(4x - 3)^2$

   (A) $16x^2 - 24x + 9$
   (B) $400x^2 - 225$
   (C) $80x - 45$
   (D) $64x^2 - 96x + 36$

   A10.0

16. Simplify: $\dfrac{4x + 7}{4x^2 - 17x - 42}$

   (A) $\dfrac{1}{x - 6}$
   (B) $\dfrac{4x + 7}{(2x - 6)(2x + 7)}$
   (C) $\dfrac{1}{4x - 6}$
   (D) Cannot be simplified

   A12.0

17. Solve for x: $7(2x + 6) - 4(9x + 6) < -26$

   (A) $x > -2$
   (B) $x > 2$
   (C) $x < -2$
   (D) $x < -1$

   A5.0

18. Which of the following is the graph of the equation $y = x + 2$?

   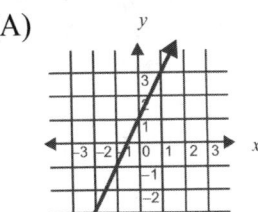

   (A)

   (B)

   (C)

   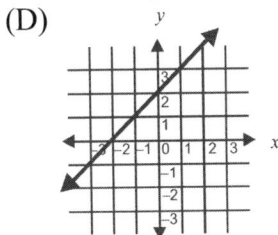

   (D)

   A6.0

19. Solve the equation $\sqrt{6w - 8} = w$.

   (A) $w = 3, 4$
   (B) $w = 2, 4$
   (C) $w = 3\sqrt{2}, 4$
   (D) $w = 2\sqrt{3}, -2\sqrt{3}$

   A14.0

20. Solve the equation $(x - 3)^2 = 1$.

   (A) $x = 3, -3$
   (B) $x = 1, 3$
   (C) $x = 2, 4$
   (D) $x = 1, -1$

   A14.0

21. Which of the following is a graph of the inequality $-y \geq 2$?

(A)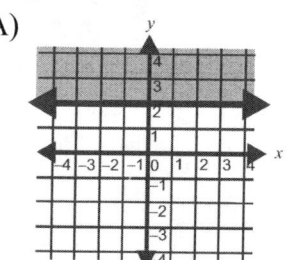

(B)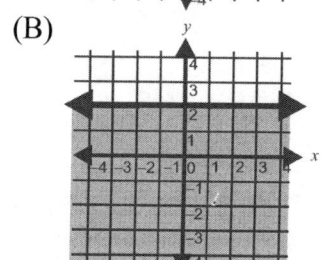

(C)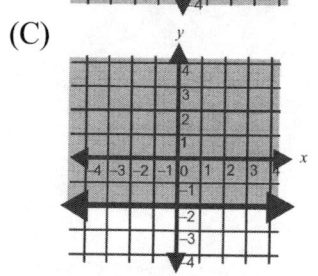

(D)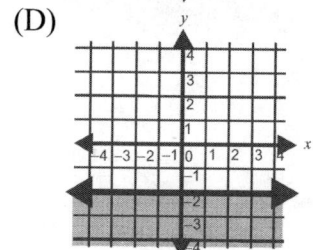

A6.0

22. The functional relationship between altitude (A) above sea level (in feet) and the approximate boiling point (B) of water (in degrees Fahrenheit) may be expressed by the equation $B = -0.00176A + 212$. What is the approximate boiling point of water at 2,500 feet?

(A) 194.4
(B) 207.6
(C) 208.3
(D) 216.4

A5.0

23. Figure $ABCDEFGH$ is a regular octagon.

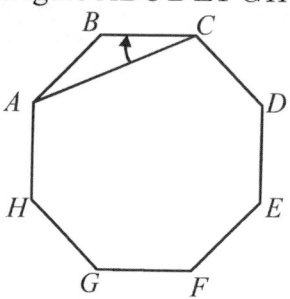

What is the measure, in degrees, of $\angle ACB$?

(A) $90°$
(B) $22.5°$
(C) $45°$
(D) $135°$

G12.0

24. Which is the graph of $2x - y = 1$?

(A)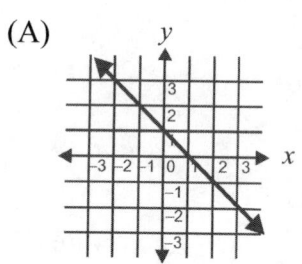

(B)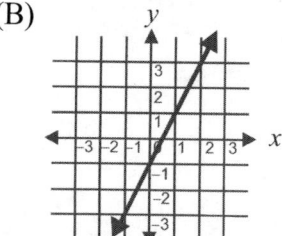

(C)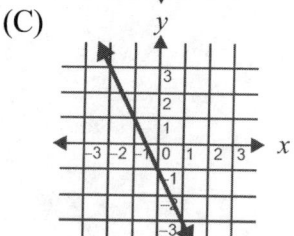

(D)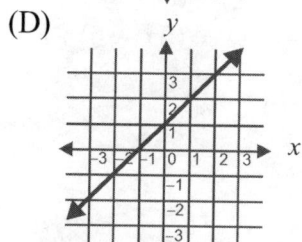

A6.0

25. What is the equation of the line that includes the point $(3, -1)$ and has a slope of 2?

   (A) $y = -2x - 7$
   (B) $y = -2x - 2$
   (C) $y = -2x + 7$
   (D) $y = 2x - 7$

   A7.0

26. Multiply: $6w(-5w^3 + 2w^2 - 4w)$

   (A) $-30w^3 + 12w^2 - 24w$
   (B) $-5w^3 + 2w^2 + 2w$
   (C) $-30w^4 + 12w^3 - 24w^2$
   (D) $-30w^4 + 12w^3 + 24w^2$

   A10.0

27. Solve the following quadratic equation by using the quadratic formula.
   $6x^2 - 16x - 6 = 0$

   (A) $2, -3$
   (B) $3, -\frac{1}{3}$
   (C) $\frac{4 \pm \sqrt{7}}{3}$
   (D) $2, -\frac{1}{3}$

   A20.0

28. Where is the y-intercept for the line $3x + 2y + 8 = 0$?

   (A) $(2, 0)$
   (B) $(0, 8)$
   (C) $(0, -4)$
   (D) $(-4, 0)$

   A6.0

29. Factor $9x^2 + 15x - 14$

   (A) $(9x - 1)(x + 7)$
   (B) $(3x + 2)(3x - 7)$
   (C) $(3x - 2)(3x - 7)$
   (D) $(3x - 2)(3x + 7)$

   A11.0

30. After the Valentine's Day party, Jenny, Luisa, and Carla had 68 pieces of candy combined. Jenny got 4 more pieces of candy than Carla, and Luisa got 6 more pieces of candy than Jenny. How many pieces did Carla get?

   (A) 22
   (B) 29
   (C) 18
   (D) 23

   A5.0

31. What are the roots of the quadratic equation below?

   $x^2 - 2x - 24 = 0$

   (A) $\{-8, 6\}$
   (B) $\{-6, 4\}$
   (C) $\{6, -4\}$
   (D) $\{-8, 3\}$

   A19.0

32. The coordinates of the endpoints of a line segments are $(3, 1)$ and $(-5, 9)$. Find the coordinates of the midpoint of the line segment.

   (A) $(4, 5)$
   (B) $(-1, 5)$
   (C) $(-2, 4)$
   (D) $(-2, 8)$

   G17.0

33. Solve by factoring: $x^2 - 5x + 6 = 0$.

   (A) -3 and -2
   (B) -3 and 2
   (C) 3 and -2
   (D) 3 and 2

   A19.0

34. Is the equation $3(2x - 6) = 12$ equivalent to the equation $6x - 18 = 12$?

   (A) Yes because of the commutative property.
   (B) Yes because of the distributive property.
   (C) Yes because of the associative property.
   (D) Yes because of the inverse property.

   A1.1

35. What is the volume of a wading pool 12 feet long, 6 feet wide, and 6 inches deep?

   (A) 18 cubic feet
   (B) 36 cubic feet
   (C) 216 cubic feet
   (D) 432 cubic feet

   G8.0

36. If you double the radius of a circle, how much does the area increase?

   (A) The area remains the same.
   (B) The area doubles.
   (C) The area is three times greater.
   (D) The area is four times greater.

   G11.0

37. Which of the following equations is represented by this graph?

   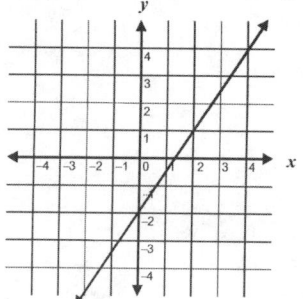

   (A) $y = 2x - 2$
   (B) $y = \frac{3}{2}x - 2$
   (C) $y = \frac{3}{2}x + 2$
   (D) $y = \frac{2}{3}x - 2$

   A7.0

38.

   What is the perimeter of the rectangle above?

   (A) 1996 m
   (B) 2000 m
   (C) 998 m
   (D) 224,665 m

   G8.0

39. Frank has purchased a cylindrical water tank with a volume of 640 cubic feet. The height of the tank is 8 feet. What is the area of the base of the tank?

   (A) 80 square feet
   (B) 251 square feet
   (C) 632 square feet
   (D) 800 square feet

   G9.0

40. A preschool is required to have a playground of at least 900 square feet. Which of the following would be satisfactory measurements for a playground for the school?

   (A) 30 feet by 32 feet
   (B) 27 feet by 30 feet
   (C) 15 feet by 40 feet
   (D) 10 feet by 80 feet

   G8.0

41. A Ferris wheel has a radius of 14 feet. How far will you travel if you take a ride that goes around six times? Use $\pi = \frac{22}{7}$.

   (A) 528 feet
   (B) 616 feet
   (C) 3,696 feet
   (D) 12,936 feet

42. Using the quadratic formula, solve the equation $d^2 - 4d + 1 = 0$.

   (A) $d = -3, -1$
   (B) $d = \sqrt{3}, 2\sqrt{3}$
   (C) $d = 2 - \sqrt{3}, \sqrt{3} + 2$
   (D) $d = 2i, -2i$

43. All carnivores are meat eaters. Lions eat meat. Therefore, Lions are carnivores. This kind of thinking is an example of _____ reasoning.

   (A) applied
   (B) inductive
   (C) qualitative
   (D) deductive

44.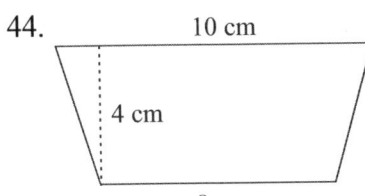
   Find the area of the trapezoid above.

   (A) 22 square centimeters
   (B) 36 square centimeters
   (C) 72 square centimeters
   (D) 320 square centimeters

45. What is the sum of the measures of the interior angles of a hexagon if each side measures 7 inches?

   (A) 360°
   (B) 540°
   (C) 720°
   (D) 960°

46. What is the sum of the measures of the interior angles of a pentagon?

   (A) 50°
   (B) 360°
   (C) 420°
   (D) 540°

47. If the measures of two sides of a triangle are 7 and 13, which of the following cannot be the measure of the third side?

   (A) 20
   (B) 19
   (C) 18
   (D) 17

48. What is the opposite of $\frac{1}{3}$?

   (A) -3
   (B) $-\frac{1}{3}$
   (C) 0.3
   (D) 3

49. What is the reciprocal of 16?

   (A) -16
   (B) $\frac{1}{16}$
   (C) 4
   (D) 256

50. Which of these graphs represents $x < -4$ or $x \geq 1$?

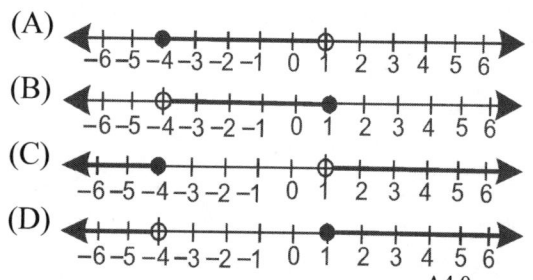

51. Matt needs to earn at least $1000 this month. He is paid $400 plus $50 for each new account he establishes. How many new accounts (a) must he establish in order to earn at least $1000?

   (A) $a \geq 8$
   (B) $a \geq 12$
   (C) $a \geq 20$
   (D) $a \geq 24$

   A5.0

52. There are 5 printers in the computer lab. Three of them each print between 5 and 7 pages per minute. The other two each print between 6 and 10 pages per minute. Which of these inequalities represents the possible numbers of pages that all five of the printers can print in 10 minutes?

   (A) 27 pages $\leq x \leq$ 41 pages
   (B) 135 pages $\leq x \leq$ 221 pages
   (C) 110 pages $\leq x \leq$ 170 pages
   (D) 270 pages $\leq x \leq$ 410 pages

   A5.0

53. $x^{\frac{1}{2}} \times x^{\frac{5}{7}} =$

   (A) $x^{\frac{6}{9}}$
   (B) $x^{\frac{5}{14}}$
   (C) $x^{\frac{17}{14}}$
   (D) $x^{\frac{5}{9}}$

   A2.0

54. Find x in the triangle below.

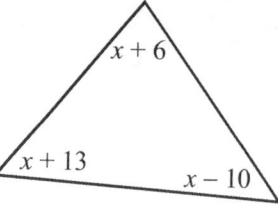

   (A) 72
   (B) 70
   (C) 63
   (D) 57

   G12.0

55. If the equation below were graphed, which of the following points would lie on the line?

   $4x + 7y = 56$

   (A) $(7, 4)$
   (B) $(0, 14)$
   (C) $(8, 0)$
   (D) $(4, 7)$

   A7.0

56. What is the area of the figure below?

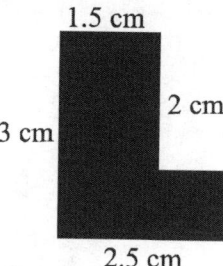

   (A) 5.5 cm^2
   (B) 9 cm
   (C) 10.5 cm^2
   (D) 11 cm

   G10.0

57. Factor $y^2 - 81$

   (A) $(y + 9)(y + 9)$
   (B) $(y - 9)(y - 9)$
   (C) $(y + 9)(y - 9)$
   (D) $(y + 3)(y - 3)$

   A11.0

58. An advertising sign is an equilateral triangle with sides 10 feet long. The area of the sign is 45 ft².

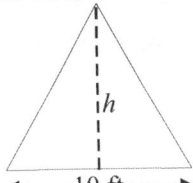

What is the height of the sign?

(A) 7 feet
(B) 18 feet
(C) 36 feet
(D) 9 feet G10.0

59. What is the area of the parallelogram?

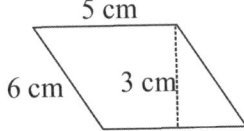

(A) 15 cm²
(B) 30 cm²
(C) 7.5 cm²
(D) 16.5 cm² G10.0

60. Find the equation of the line that passes through the points $(2, 3)$ and $(-2, 1)$.

(A) $y = -\frac{1}{2}x + 4$
(B) $y = 2x - 1$
(C) $y = \frac{1}{2}x + 2$
(D) $y = -2x + 7$ A7.0

61. Simplify: $\dfrac{a^2 - 9}{a^2 - 4a + 3}$

(A) $\dfrac{a-3}{a-1}$
(B) $\dfrac{(a+3)(a-3)}{(a-3)(a-1)}$
(C) $\dfrac{a+3}{a-1}$
(D) $\dfrac{a+3}{a+1}$ A12.0

62. Simplify: $\dfrac{2x^2 - 4x - 6}{x - 3}$

(A) $\dfrac{2(x-3)(x+1)}{x-3}$
(B) $\dfrac{(2x-6)(x+1)}{x-3}$
(C) $2(x+1)$
(D) $x - 3$ A12.0

63. Simplify: $\dfrac{5}{y} - \dfrac{2}{y}$

(A) $\dfrac{7}{y^2}$
(B) $\dfrac{3}{y}$
(C) $\dfrac{7}{y}$
(D) $\dfrac{3}{y^2}$ A13.0

64. Simplify: $\dfrac{x}{y} \times \dfrac{y}{x}$

(A) $\dfrac{1}{x}$
(B) $\dfrac{1}{y}$
(C) 1
(D) x A13.0

65. Simplify: $\dfrac{x-2}{x+7} \div \dfrac{x-7}{x+2}$

(A) $\dfrac{x^2 - 9x + 14}{x^2 + 9x + 14}$
(B) $\dfrac{x^2 - 9x + 7}{x^2 + 9x + 7}$
(C) $\dfrac{x-7}{x+7}$
(D) $\dfrac{x^2 - 4}{x^2 - 49}$ A13.0

Practice Test 2

1. Add: $\dfrac{3x-2}{x+1} + \dfrac{x^2+x-2}{x^2-1}$

   (A) $\dfrac{x^2+4x-4}{x^2+x}$

   (B) $\dfrac{x^2-2x-4}{x^2-x-2}$

   (C) $\dfrac{4x}{x+1}$

   (D) $\dfrac{3x^3}{x+1}$

   A13.0

2. What is the slope of a line perpendicular to the line passing through the points $(3,6)$ and $(5,1)$?

   (A) $-\dfrac{5}{2}$

   (B) $-\dfrac{4}{3}$

   (C) $-\dfrac{3}{4}$

   (D) $\dfrac{2}{5}$

   A8.0

3. What is the reciprocal of -52?

   (A) $-\dfrac{1}{52}$

   (B) $\dfrac{1}{52}$

   (C) 26

   (D) 52

   A2.0

4. $14(x-6) = -26$

   (A) $x = 58$
   (B) $x = 4\tfrac{1}{7}$
   (C) $x = -7\tfrac{6}{7}$
   (D) $x = 29$

   A4.0

5. Find the area of the following trapezoid.

   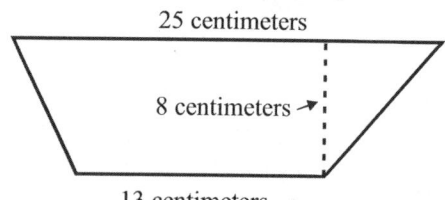

   (A) 200 square centimeters
   (B) 104 square centimeters
   (C) 152 square centimeters
   (D) 304 square centimeters

   G10.0

6. Subtract: $\dfrac{x+1}{x} - \dfrac{2x-4}{x}$

   (A) $\dfrac{3x-5}{x}$

   (B) $\dfrac{3x-3}{x^2}$

   (C) $\dfrac{5-x}{x}$

   (D) $\dfrac{x-5}{x}$

   A13.0

7. Multiply: $\dfrac{ab}{c} \times \dfrac{a}{b}$

   (A) $\dfrac{a^2}{bc}$

   (B) $\dfrac{a^2}{c}$

   (C) $\dfrac{a}{bc}$

   (D) $\dfrac{a^2}{b}$

   A13.0

8. $4^{-2} \times 2^{-3}$

   (A) -48

   (B) $\dfrac{1}{128}$

   (C) $\dfrac{1}{48}$

   (D) $\dfrac{3}{36}$

   A2.0

9. Four students attempt to simplify a mathematical expression. They have four different answers. Which of the answers below is equivalent to the expression,

   $2(a + 3b) - 4(3a - b) - (5a + 4b)$?

   (A) $-15a + 6b$

   (B) $-17a + 9b$

   (C) $-9a - 2b$

   (D) $-9a + 9b$

   A10.0

10. Use the correct order of operations to evaluate the following expression.

    $-3(x - 5)^2$

    (A) $-3x^2 - 10x + 25$

    (B) $x^2 - 10x + 25$

    (C) $3x^2 - 30x + 75$

    (D) $-3x^2 + 30x - 75$

    A10.0

11. Simplify the following monomial.

    $2 \cdot x^4 \cdot y^6 \cdot x^{-4}$

    (A) $2y^6$

    (B) $2(xy)^6$

    (C) $64y^6$

    (D) $2x^{-8}y^6$

    A2.0

12. Solve for x in the following equation.

    $\dfrac{6x - 19}{-2} = 3.5$

    (A) 12

    (B) $\dfrac{13}{3}$

    (C) 2

    (D) 4

    A2.0

13. Solve: $3(5x + 3) + 5(4x - 9) = 34$

    (A) $x = 1$

    (B) $x = 2$

    (C) $x = -1$

    (D) $x = -2$

    A4.0

14. Solve $-4(2x + 7) > 3(4x + 5) + 27$

    (A) $x > \dfrac{7}{2}$

    (B) $x < \dfrac{7}{2}$

    (C) $x < -\dfrac{7}{2}$

    (D) $x > \dfrac{1}{4}$

    A4.0

15. If the equation below were graphed, which of the following points would lie on the line?

    $x - 7y = 21$

    (A) $(7, 3)$

    (B) $(0, -3)$

    (C) $(14, 0)$

    (D) $(-3, 14)$

    A7.0

16. Find the x- and y- intercepts for the following equation: $2x + 5y = 30$.

   (A) x-intercept = 15
       y-intercept = 6
   (B) x-intercept = 5
       y-intercept = 4
   (C) x-intercept = 6
       y-intercept = 15
   (D) x-intercept = 4
       y-intercept = 5

   A6.0

17. What are the x and y intercepts of the equation graphed below?

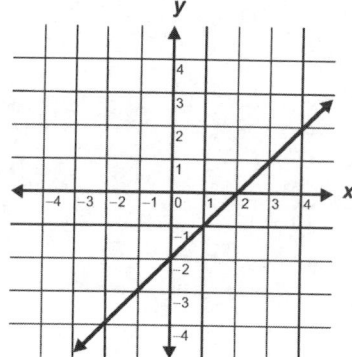

   (A) x-intercept = 1; y-intercept = -1
   (B) x-intercept = -2; y-intercept = 2
   (C) x-intercept = -1; y-intercept = 1
   (D) x-intercept = 2; y-intercept = -2

   A6.0

18. Simplify: $3(5x - 2) + (-4x + 5)$

   (A) $4x$
   (B) $4x - 7$
   (C) $11x - 11$
   (D) $11x - 1$

   A10.0

19. Which of these graphs represents the inequality $y \geq 2x + 1$?

   (A)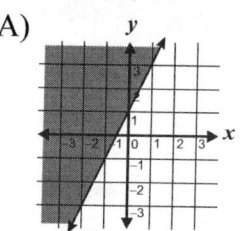

   (B)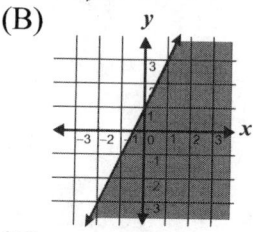

   (C)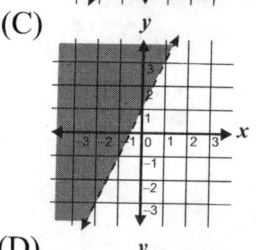

   (D)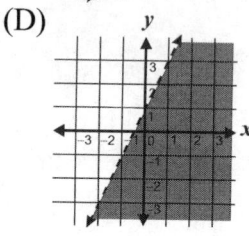

   A6.0

20. Berny and Harvey are spending a week visiting their friend Wayne who lives on a ranch. The first morning of their visit they are awakened by a rooster crowing. The second morning they are again awakened by a rooster crowing. Again, on the third morning, sure enough, the rooster crows again. Berny, who has studied logic at school, says "Harvey, I bet that rooster will crow and wake us up again tomorrow morning." Berny's conjecture is based on

   (A) empirical hypothesizing.
   (B) deductive reasoning.
   (C) linear extrapolation.
   (D) inductive reasoning.

   A25.1

21. Which of these graphs represents $y = -\frac{1}{3}x - 2$?

(A)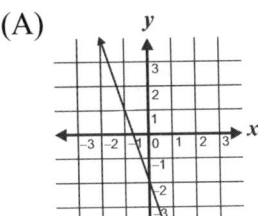

(B)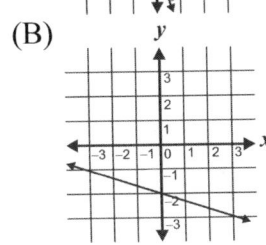

(C)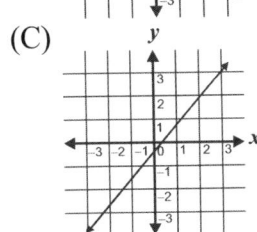

(D)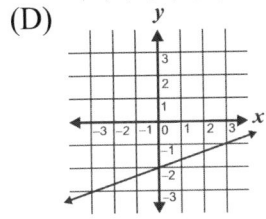

A6.0

22. Solve the equation $(x+9)^2 = 49$

   (A) $x = -9, 9$
   (B) $x = -9, 7$
   (C) $x = -16, -2$
   (D) $x = -7, 7$

   A14.0

23. Solve using the quadratic formula: $6a^2 + 11a - 10 = 0$

   (A) $\{-\frac{2}{5}, \frac{3}{2}\}$
   (B) $\{\frac{2}{5}, \frac{2}{3}\}$
   (C) $\{-\frac{5}{2}, \frac{2}{3}\}$
   (D) $\{\frac{5}{2}, \frac{2}{3}\}$

   A20.0

24. Solve the equation using the quadratic formula: $y = \sqrt{16 - 6y}$

   (A) $y = 4, 2\sqrt{3}$
   (B) $y = 3 - \sqrt{2}, 2 - \sqrt{3}$
   (C) $y = 2, -8$
   (D) $y = 2 + 3i, 2 - 3i$

   A20.0

25. Solve the equation $c^2 + 3c - 9 = 0$ by completing the square.

   (A) $c = 3, -3$
   (B) $c = \frac{3}{2}\sqrt{5} - \frac{3}{2}, -\frac{3}{2}\sqrt{5} - \frac{3}{2}$
   (C) $c = \pm\sqrt{3}$
   (D) $c = 3i, -3i$

   A14.0

26. Multiply: $-6a^3(-2ab^2 + 5a^2b - 6a^3)$

   (A) $12a^3b^2 - 30a^6 + 36a^9$
   (B) $12a^4b^2 - 30a^5b + 36a^6$
   (C) $-12a^3b^2 + 30a^6 - 36a^9$
   (D) $-12a^4b^2 + 30a^5b - 36a^6$

   A10.0

27. Solve: $-\frac{4}{5}x \geq 8$

   (A) $x \geq 10$
   (B) $x \geq 5$
   (C) $x \leq -10$
   (D) $x \geq 10$

   A5.0

28. Solve for x in the following equation: $x + 2(x + 200) + 800 = 3000$

   (A) $1,200$
   (B) 800
   (C) 600
   (D) $1,600$

   A5.0

29. Factor: $b^2 - 2b - 8$

   (A) $(b-4)(b+4)$
   (B) $(b-2)(b+4)$
   (C) $(b+2)(b-4)$
   (D) $(b-2)(b-2)$

   A11.0

30. Al weighs 5 pounds less than three times Little Bill's weight. Which equation represents this statement?

   (A) $a - 5 = 3b$
   (B) $a - 5 = b - 3$
   (C) $a = 5 - 3b$
   (D) $a = 3b - 5$

   A5.0

31. The coordinates for the endpoints of a line segment are $(-5, 13)$ and $(-9, 21)$. Find the coordinates for the midpoint of the line segment.

   (A) $(-7, 17)$
   (B) $(-4, 4)$
   (C) $(-2, 4)$
   (D) $(4, 17)$

   G17.0

32. Factor: $4x^2 + x - 3$

   (A) $(2x+3)(2x-1)$
   (B) $(4x-3)(x+1)$
   (C) $(2x-3)(2x+1)$
   (D) $(4x+3)(x-1)$

   A11.0

33. Solve the following equation: $3x^2 = 9x$

   (A) $x = \{0, 1\}$
   (B) $x = \{3, 1\}$
   (C) $x = \{0, 3\}$
   (D) $x = \{3, -3\}$

   A14.0

34. Solve: $4y^2 - 9y = -5$

   (A) $\left\{1, \frac{5}{4}\right\}$
   (B) $\left\{-\frac{3}{4}, -1\right\}$
   (C) $\left\{-1, \frac{4}{5}\right\}$
   (D) $\left\{\frac{5}{16}, 1\right\}$

   A19.0

35. Solve for y: $2y^2 + 13y + 15 = 0$

   (A) $\left\{\frac{3}{2}, \frac{5}{2}\right\}$
   (B) $\left\{\frac{2}{3}, \frac{2}{5}\right\}$
   (C) $\left\{-5, -\frac{3}{2}\right\}$
   (D) $\left\{5, -\frac{3}{2}\right\}$

   A19.0

36. The coordinates of a line segment are $(1, 6)$ and $(11, -4)$. What are the coordinates for the midpoint?

   (A) $(6, 1)$
   (B) $(10, 2)$
   (C) $(5, 1)$
   (D) $(12, 2)$

   G17.0

37. Yon works for Zap Electric as an electrician. When he is sent out on a job, he is told to charge a $120 travel fee + $80/hour. Yon uses the formula $c = 80h + 120$. Yon worked 4 hours on a job. How much did he charge?

   (A) $200
   (B) $204
   (C) $400
   (D) $440

   A5.0

38. The living room in Ty's house has 168 square feet of floor space. His family is building an addition to this room that measures 14 feet long and 8 feet wide. What will be the total square feet of the living room with the new addition?

   (A) 112 square feet
   (B) 180 square feet
   (C) 270 square feet
   (D) 280 square feet

   G10.0

39. Divide: $\dfrac{x}{y} \div \dfrac{y}{x}$

   (A) $\dfrac{x^2}{y^2}$
   (B) 1
   (C) $\dfrac{xy}{x}$
   (D) $\dfrac{xy}{y}$

   A13.0

40. If you have a 6 inch cube and decide when you make the next cube, you are going to double the length of each side, how will the volume be affected?

   (A) The volume will be 3 times larger.
   (B) The volume will be twice as large.
   (C) The volume will be 8 times larger.
   (D) The volume will be 9 times larger.

   G11.0

41. Find the equation of the line that contains the points $(-2, -3)$ and $(0, 4)$.

   (A) $y = \tfrac{2}{7}x - 4$
   (B) $y = \tfrac{7}{2}x + 4$
   (C) $y = -\tfrac{2}{7}x - 3$
   (D) $y = -\tfrac{7}{2}x - 3$

   A7.0

42. What is the volume of the box shown below?

   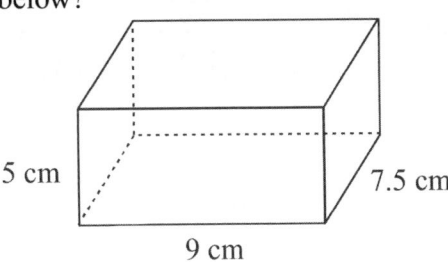

   (A) 22.5 cm³
   (B) 337.5 cm³
   (C) 1875 cm³
   (D) 2250 cm³

   G9.0

43. What is the area of the triangle?

   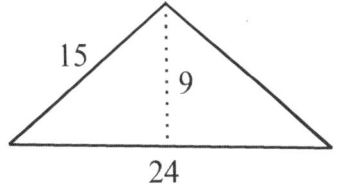

   (A) 54 square units
   (B) 67.5 square units
   (C) 108 square units
   (D) 216 square units

   G10.0

44. What is the surface area of the rectangular prism with a width of 8 in, a length of 5 inches, and a height of 4 in?

   (A) 34 in²
   (B) 92 in²
   (C) 160 in²
   (D) 184 in²

   G8.0

45. The sum of the three angles in a triangle equals

   (A) 90°.
   (B) 180°.
   (C) 360°.
   (D) 540°.

   G12.0

46. Frank works weekends at a grocery store. He works 5 hours on Saturday and 5 hours on Sunday. He is paid $10 per hour. He claims this is approximately $5000 per year. Is his claim reasonable?

(A) Yes
(B) No, he makes much less than $5,000 per year.
(C) No, he makes much more than $5,000 per year.
(D) There is not enough information to evaluate the reasonableness of his claim.

A25.2

47. Find the perimeter of the trapezoid.

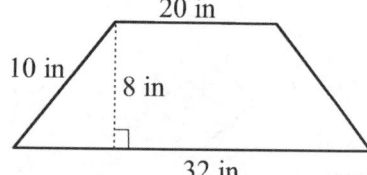

(A) 70 in
(B) 72 in
(C) 208 in
(D) 860 in

G8.0

48. The coordinates of a line segment are $(1, 6)$ and $(11, -4)$. What is the length of the line segment? Round your answer to the nearest tenth.

(A) 7.1
(B) 14.1
(C) 50
(D) 200

G17.0

49. The area of a circle with radius 98 mm is approximately

(A) 308 mm^2.
(B) 616 mm^2.
(C) 7543 mm^2.
(D) 30156 mm^2.

G8.0

50. Alfred calculated the volume of the water tank to be approximately 150 cubic feet. He knew this was wrong because

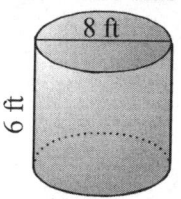

(A) $6 \times 6 + 8 \times 8 = 100$
(B) $3 \times 4 \times 4 \times 6 = 288$
(C) $8 \times 8 \times 6 = 384$
(D) $8 \times 8 \times 6 \times 6 = 2304$

G8.0

51. What is the area of the parallelogram?

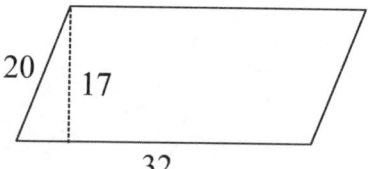

(A) 272 square units
(B) 320 square units
(C) 544 square units
(D) 640 square units

G10.0

52. What is the volume, in cubic feet, of the square pyramid below?

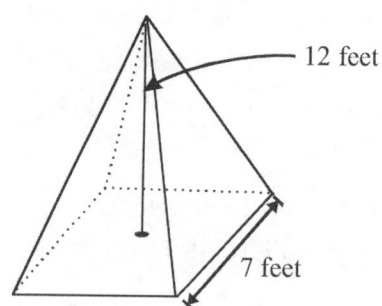

(A) 168 cubic feet
(B) 196 cubic feet
(C) 294 cubic feet
(D) 588 cubic feet

G9.0

53. Using the polygon angle sum theorem, what is the measure of x?

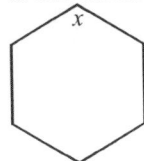

(A) 60°
(B) 120°
(C) 180°
(D) 720°

G12.0

54. Mary has budgeted $50 to rent a rototiller to prepare a garden. The charge is $15 for the first hour, and $5 per hour for each additional hour. For how many hours can she use the rototiller?

(A) $3\frac{1}{3}$ hr
(B) 4 hr
(C) 7 hr
(D) 8 hr

A5.0

55. The perimeter of a rectangle is 160 feet. The length of the rectangle is 20 feet less than three times the width. What is the width and length of the rectangle?

(A) $w = 15, l = 25$
(B) $w = 25, l = 15$
(C) $w = 30, l = 15$
(D) $w = 25, l = 55$

A5.0

56. What is the equation of a line that has a slope of 0 and goes through the point $(2, 3)$?

(A) $x = 2$
(B) $x = 3$
(C) $y = 2$
(D) $y = 3$

A7.0

57. Using the Triangle Inequality Theorem, which inequality is true about the triangle below?

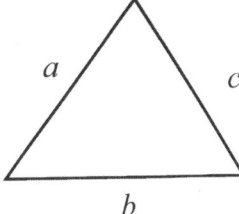

(A) $a + b > c$
(B) $b + c > a$
(C) $a + c > b$
(D) All of the above are true.

G6.0

58. Simplify: $\dfrac{4a^2b}{16ab^3}$

(A) $\dfrac{a^2b}{4ab^3}$
(B) $\dfrac{1}{4ab^2}$
(C) $\dfrac{ab}{4b^3}$
(D) $\dfrac{a}{4b^2}$

A12.0

59. Divide $4x^3y - 5x^2 + y^2$ by $2x$ and simplify.

(A) $\dfrac{4x^3y - 5x^2 + y^2}{2x}$
(B) $2x^2y - 5x^2 + y^2$
(C) $2x^2y - \dfrac{5x}{2} + \dfrac{y^2}{2x}$
(D) $4x^2y - \dfrac{5x}{2} + y^2$

A12.0

60. Which of the following equations is represented by this graph?

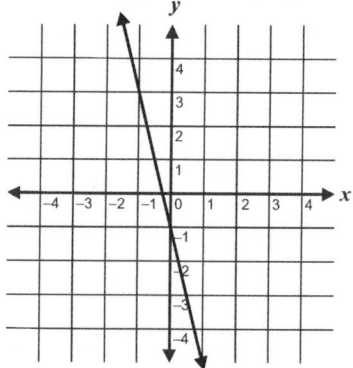

(A) $y = 4x - 1$

(B) $y = -4x - 1$

(C) $y = \frac{1}{4}x - 1$

(D) $y = -\frac{1}{4}x - 1$

A7.0

61. Simplify $\frac{2a^3bc}{9ab^4c^2}$

(A) $\frac{2a^2}{9b^3c}$

(B) $\frac{2a^3}{9b^4c^2}$

(C) $\frac{2a^2c}{9b^3c^2}$

(D) $\frac{2a^2}{9b^3}$

A12.0

62. What is the quadratic formula?

(A) $a^2 + b^2 = c^2$

(B) $ax^2 + bx + c = 0$

(C) $\frac{-b \pm \sqrt{2b - ac}}{a}$

(D) $\frac{-b \pm \sqrt{b^2 - 4ac}}{2a}$

A20.0

63. What are the measures of the three angles on the triangle?

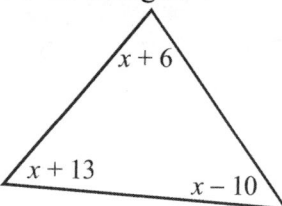

(A) 47°, 60°, 63°
(B) 47°, 57°, 60°
(C) 57°, 60°, 63°
(D) 47°, 63°, 70°

G12.0

64. What is the measure of y?

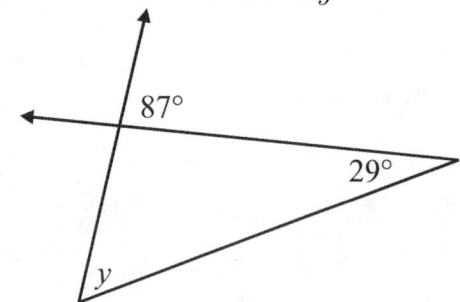

(A) 93°
(B) 58°
(C) 87°
(D) 70°

G12.0

65. What is the measure of x?

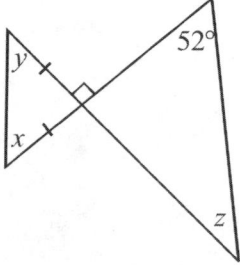

(A) 38°
(B) 45°
(C) 52°
(D) Cannot be determined

G12.0

Index

Acknowledgements, ii
Addition
  of polynomials, 68
Algebra
  consecutive integer problems, 63
  multi-step problems, 52
  two step problems, 46
      with fractions, 47
  vocabulary, 34
  word problems, 36, 58
      age problems, 60
      setting up, 38
Angles
  sum of polygon interior, 148
Area
  circle, 155
  rhombus, 153
  squares and rectangles, 150
  triangles, 151
  two-step problems, 156
Arguments, 135
  inductive and deductive, 137
Associative Property
  of addition, 41
  of multiplication, 41

Base, 34
Binomials, 66, 86
  multiplication
      FOIL method, 76

Cartesian Plane, 112
Circumference, Circle, 154
Coefficient, 34
  leading, 34
Combining like terms, 49
Commutative Property
  of addition, 41
  of multiplication, 41

Completing the Square, 106
Conclusion, 135
Constant, 34
Contrapositive, 138
Converse, 138
Counterexample, 138

Deductive Reasoning, 135
Degree, 34
Denominator, 23, 74
  dividing monomials, 72
Diagnostic Test, 1
Diameter, Circle, 154
Difference of Two Squares, 103
Distance Formula
$$d = \sqrt{(y_2 - y_1)^2 + (x_2 - x_1)^2}, 115$$
Distributive Property, 41

Equations
  finding using two points or a point and slope, 123
  linear, 111, 112
  of perpendicular lines, 124
  solving with like terms, 49
Equilateral Triangle, 141
Evaluation Chart, 11
Exponents, 66
  dividing, 20, 21
  dividing monomials, 72
  multiplication using, 15, 16
  multiplying polynomials, 70
  negative, 19
  of monomials with different variables, 71
  of polynomials, 67
  simpifying monomial expressions, 78
  when subtracting polynomials, 68

Factoring
  by grouping, 85
  difference of two squares, 91
  of polynomials, 82

quadratic equations, 102
trinomials, 86
FOIL method, 76, 85, 86
for multiplying binomials, *see* Binomials
Fractions
monomials, 72
raised to a power, 17

Geometric Relationships of Plane Figures, 158
Geometry
word problems, 59
Graphing
a line knowing a point and slope, 122
horizontal and vertical lines, 114
inequalities, 127
linear equations, 111, 112

Identity Property
of multiplication, 41
of zero, 41
Inductive reasoning, 135
Inequalities
multi-step, 54
word problems, 63
Inequality
definition, 34
Integers, 70
Intercepts of a Line, 117
Inverse, 138
Inverse Property
of addition, 41
of multiplication, 41
Irrational Numbers, 12
Isosceles Triangle, 141

Linear Equation, 112, 120
Linear equation, 111
general form, 111
slope-intercept form, 111
standard form, 111
Lines
parallel, 124
Logic, 135

Mathematical Reasoning, 135
Midpoint of a Line Segment
$M = \left(\frac{x_1+x_2}{2}, \frac{y_1+y_2}{2}\right)$, 116
Monomials, 66
adding and subtracting, 66
dividing, *see* Fractions
multiplying, 70, 71
multiplying by polynomials, 73
multiplying more than two, 71
Multi-Step Algebra Problems, 52

Numerator, 23
dividing polynomials, 72

Opposites, 13
Order of Operations, 22
Ordered Pair, 112

Parallel lines, 124
Parentheses
removing, 51
Parentheses, removing and simplifying
polynomials, 75
Perfect Squares, 91, 105
Perimeter, 59, 149
Perpendicular Lines
equations of, 124
π, 154
Point-Slope Form of an Equation
$y - y_1 = m(x - x_1)$, 123
Polygons, 147, 149
sum of interior angles, 148
Polynomial(s), 66, 82
adding, 67
dividing by monomials, 74
factoring, 82
greatest common factor, 82–84
multiplying by monomials, 73
subtracting, 68
Practice Test 1, 179
Practice Test 2, 188
Preface, viii
Premises

of an argument, 135
Product, 82
Proposition, 135

Quadratic equation, 101
   $ax^2 + bx + c = 0$, 108
Quadratic formula
   $\dfrac{-b \pm \sqrt{b^2 - 4ac}}{2a}$, 108

Radius, Circle, 154
Rational Expressions
   adding, 94
   dividing, 97
   multiplying, 96
   simplifying, 93
   subtracting, 95
Rational Numbers, 12
Real Numbers, 12
Reciprical, 14
Rectangle, 59
Reflexive Property of Equality, 41
Right Triangle, 141
Roots
   adding and subtracting, 28
   dividing, 30
   multiplying, 29

Sentence, 34
Slope, 120
   $m = \dfrac{y_2 - y_1}{x_2 - x_1}$, 118
Slope Intercept Form of a Line
   $y = mx + b$, 120
Square root
   simplifying, 27
Square Roots, 27
Substitution
   numbers for variables, 35
Subtraction
   of polynomials, 68

Subtrahend, 68
Surface Area
   cone, 173
   cube, 169
   cylinder, 172
   pyramid, 171
   rectangular prism, 169
   sphere, 173
Symmetric Property of Equality, 41

Table of Contents, vii
Term, 34
Transitive Property of Equality, 41
Triangle, 59
   isosceles, 60
Triangle Inequality Theorem, 143
Triangles
   equilateral, 141
   exterior angles, 143
   isosceles, 141
   right, 141
Trinomials, 66
   factoring, 86
Two Step Algebra Problems, 46
   with fractions, 47
Two-Step Area Problems, 156
Two-Step Volume Problems, 166

Variable, 34, 50, 66, 67, 70
Volume
   rectangular prisms, 163
   rectangular solids, 162
   spheres, cones, cylinders, and pyramids, 164
   two-step problems, 166

Word Problems
   algebra, 36
      setting up, 38
   changing to algebraic equations, 39
   geometry, 59

TEMESCAL CANYON HIGH SCHOOL
LIBRARY MEDIA CENTER
28755 EL TORO ROAD
LAKE ELSINORE, CA. 92532